UK Standard Industrial Classification of Economic Activities 2007 (SIC 2007)

Structure and explanatory notes

Editor: Lindsay Prosser
Office for National Statistics

palgrave macmillan

ISBN 978-0-230-21012-7

A National Statistics publication

National Statistics are produced to high professional standards as set out in the Code of Practice for Official Statistics. They are produced free from political influence.

About us

The Office for National Statistics

The Office for National Statistics (ONS) is the executive office of the UK Statistics Authority, a non-ministerial department which reports directly to Parliament. ONS is the UK government's single largest statistical producer. It compiles information about the UK's society and economy, and provides the evidence-base for policy and decision-making, the allocation of resources, and public accountability. The Director-General of ONS reports directly to the National Statistician who is the Authority's Chief Executive and the Head of the Government Statistical Service.

The Government Statistical Service

The Government Statistical Service (GSS) is a network of professional statisticians and their staff operating both within the Office for National Statistics and across more than 30 other government departments and agencies.

Palgrave Macmillan

This publication first published 2010 by Palgrave Macmillan.

Palgrave Macmillan in the UK is an imprint of Macmillan Publishers Limited, registered in England, company number 785998, of Houndmills, Basingstoke, Hampshire RG21 6XS.

Palgrave Macmillan in the US is a division of St Martin's Press LLC, 175 Fifth Avenue, New York, NY 10010.

Palgrave Macmillan is the global academic imprint of the above companies and has companies and representatives throughout the world.

Palgrave® and Macmillan® are registered trademarks in the United States, the United Kingdom, Europe and other countries.

A catalogue record for this book is available from the British Library.

10   9   8   7   6   5   4   3   2   1
18 17 16 15 14 13 12 11 10 09

Contacts

This publication

For information about the content of this publication, contact the Editor

Tel: 01329 446389
Email: lindsay.robinson@ons.gsi.gov.uk

Other customer enquiries

ONS Customer Contact Centre
Tel: 0845 601 3034
International: +44 (0)845 601 3034
Minicom: 01633 812399
Email: info@statistics.gsi.gov.uk
Fax: 01633 652747
Post: Room 1015, Government Buildings,
Cardiff Road, Newport, South Wales NP10 8XG
www.ons.gov.uk

Media enquiries

Tel: 0845 604 1858
Email: press.office@ons.gsi.gov.uk

Publication orders

To obtain the print version of this publication, contact Palgrave Macmillan
Tel: 01256 302611
www.palgrave.com/ons
Price: £55.00

Copyright and reproduction

Printing

This book is printed on paper suitable for recycling and made from fully managed and sustained forest sources. Logging, pulping and manufacturing processes are expected to conform to the environmental regulations of the country of origin.

Printed and bound in Great Britain by Hobbs the Printer Ltd, Totton, Southampton

Typeset by Academic + Technical Typesetting, Bristol

Contents

Preface

This publication represents a major revision of the UK Standard Industrial Classification of Economic Activities (SIC), commenced in 2002 and completed in 2007. It is effective from 1 January 2008 and is the outcome of a series of consultations carried out in conjunction with a major revision of the European Union's industrial classification system, NACE (NACE Rev. 2).

The consultations involved many stakeholders:

- the national statistical institutes of all EU member states

- the European Commission

- In the UK, a range of government departments, the Bank of England, the devolved administrations, business and trade associations and other interested bodies

- European business and trade associations

The UK is required by European legislation to revise the SIC in parallel with NACE so that both systems remain identical down to and including the four digit class level. A further breakdown is provided for certain classes by the addition of a five digit subclass level. Both the UK SIC (2007) and NACE Rev. 2 are completely consistent with the fourth revision of the UN's International Standard Industrial Classification of all Economic Activities (ISIC Rev. 4).

These revisions are motivated by the need to adapt the classifications to changes in the world economy. The revised classifications reflect the growing importance of service activities in the economy over the last 15 years, mainly due to the developments in information and communication technologies (ICT).

The advance copy of the UK SIC (2007) structure and explanatory notes published on the website on 29 January 2007 has undergone minor revision and forms the main part of this publication.

Contact

Office for National Statistics
Classifications and Harmonisation Unit
Segensworth Road
Titchfield
Fareham
Hampshire
PO15 5RR

Classifications Helpdesk

Tel: 01329 444970
Email: classifications.helpdesk@ons.gsi.gov.uk

December 2009

Introduction

A Standard Industrial Classification (SIC) was first introduced into the UK in 1948 for use in classifying business establishments and other statistical units by the type of economic activity in which they are engaged. The classification provides a framework for the collection, tabulation, presentation and analysis of data, and its use promotes uniformity. In addition, it can be used for administrative purposes and by non-government bodies as a convenient way of classifying industrial activities into a common structure.

Classification changes

Since 1948 the classification has been revised in 1958, 1968, 1980, 1992, 1997, and 2003. Revision is necessary because, over a period of time, new products and new industries to produce them emerge, and shifts of emphasis occur in existing industries. It is not always possible for the system to accommodate such developments and, after a period of time, updating the classification is the most sensible action. The 1997 and 2003 changes were not full-scale revisions but responses to user demand for additional detail at the subclass level together with some minor renumbering and revisions. This latest publication is a major revision reflecting contemporaneous changes in NACE (see next paragraph).

The need for change equally effects all international classifications and they are revised from time to time to bring them up to date. On 9 October 1990 the European Council of Ministers passed a regulation to introduce a new statistical classification of economic activities in the European Communities (NACE Rev. 1). In January 2003, a minor revision of NACE Rev. 1, NACE Rev. 1.1, was published[1] followed by a major revision, NACE Rev. 2, effective from 1 January 2008[2].

International classifications

From the outset, the UK SIC followed the same broad principles as the relevant international systems. UK statisticians played an important part in the formulation of the first ISIC (International Standard Industrial Classification of All Economic Activities), issued by the United Nations in 1948 and revised in 1958, 1968, 1989, 2003 and now in 2008. Nevertheless, there were differences in detail between the two as ISIC reflected the structure of economic activity in the world as a whole rather than that in one particular country.

In 1980, one of the principal objectives of the revision of the SIC was to examine and eliminate differences from the activity classification issued by the Statistical Office of the European Communities (Eurostat) and entitled *Nomenclature générale des activités économiques dans les Communautés européennes*, usually abbreviated to NACE. This 1970 NACE could be rearranged to agree with ISIC at aggregated levels but departed from it in the details. The 1980 revision of the SIC applied NACE as closely as was practicable to the structure of British industry.

In 1990, however, the first revision of NACE was made by EC regulation and this presented a different set of circumstances.

EC regulation

A European Community regulation is directly applicable in all member states. It does not have to be confirmed by national parliaments in order to have binding effect. The NACE regulation, therefore, made it obligatory on the UK to introduce a new *Standard Industrial Classification, SIC (92)*, based on NACE Rev. 1, and to use it where the UK is required to transmit to the European Commission statistics on economic activity.

The NACE regulation gives effect to the wish of Eurostat to establish a common statistical classification of economic activities in order to promote comparability between national and Community classifications and, therefore, between national and Community statistics. The regulation applies to the use of NACE for statistical purposes only, although a country can also use NACE for administrative purposes. The regulation does not oblige member states to collect, publish or supply data. NACE is only a language and all requests for data collection, transmission and publication must be specified elsewhere. As noted already, there is now a new version of NACE, NACE Rev. 2, effective from 1 January 2008.

As already indicated, NACE was originally an acronym but now all countries use 'NACE' to describe the European Community classification of Economic Activities.

UK SIC structure

The UK SIC is based exactly on NACE but, where it was thought necessary or helpful, a fifth digit has been added to form subclasses of the NACE four digit classes. Thus, the UK SIC is a hierarchical five digit system. UK SIC (2007) is divided into 21 sections, each denoted by a single letter from A to U. The letters of the sections can be uniquely defined by the next breakdown, the divisions (denoted by two digits). The divisions are then broken down into groups (three digits), then into classes (four digits) and, in several cases, again into subclasses (five digits). So for example we have:

section C   Manufacturing (comprising divisions 10 to 33)

division 13   Manufacture of textiles

group 13.9   Manufacture of other textiles

class 13.93   Manufacture of carpets and rugs

subclass 13.93/1   Manufacture of woven or tufted carpets and rugs

There are 21 sections, 88 divisions, 272 groups, 615 classes and 191 subclasses. The full structure of UK SIC (2007) is shown on pages 27 to 53.

As with SIC (2003), the full number of arithmetically possible subdivisions at each level is not necessarily created, varying according to the diversity of activities. For example, section A (Agriculture, forestry and fishing), has only three divisions, 01, 02 and 03 whereas section C (Manufacturing) is divided into 24 divisions. The use of 0 as the final digit for a group or class indicates that it is the only subdivision at that level. Thus division 36 (Water collection, treatment and supply) has only one group, 36.0 and only one class, 36.00. On the other hand, division 49 (Land transport and transport via pipelines) has five groups and so is numbered 49.1 through to 49.5.

Links to international classifications

To the four digit level, UK SIC (2007) follows NACE Rev. 2 exactly. The difference is in the UK SIC (2007) subclasses. For example, in both NACE Rev. 2 and UK SIC (2007) , class 56.10 represents 'Restaurants and mobile food service activities'. In UK SIC (2007) , however, three subclasses are added so that 56.10/1 is 'Licensed restaurants', 56.10/2 is 'Unlicensed restaurants and cafes' and 56.10/3 is 'Take away food shops and mobile food stands'. The first two levels (sections and divisions) of UK SIC (2007) and NACE Rev. 2 are exactly the same as in ISIC Rev..4 in content and in coding. Beyond this, and shown after a decimal point in the SIC and NACE codes, the three digit groups and the four digit classes can be directly converted to the ISIC headings or can be combined to reach the ISIC heading but may have different code numbers. For example, ISIC class 8521 = SIC/NACE class 85.31 whilst ISIC class 5813 = SIC/NACE classes 58.13 + 58.14). The aim of the further breakdowns in NACE Rev. 2 and UK SIC (2007) is to obtain classifications more suited to the European and UK economies.

Changes from SIC (2003) to SIC (2007)

While some of the rules for application of the SIC have been changed, and criteria for construction of the classification, as well as the formulation of explanatory notes, have been reviewed, the overall characteristics of the SIC remain unchanged.

New concepts at the highest level of the classification have been introduced, and new detail has been created to reflect different forms of production and emerging new industries. At the same time, efforts have been made to maintain the structure of the classification in all areas that do not explicitly require change based on new concepts.

The detail of the classification has substantially increased (from 514 to 615 classes). For service-producing activities, this increase is visible at all levels, including the highest one, while for other activities, such as agriculture, the increase in detail affected mostly the lower level of the classification. One effect of the increase in detail at class level has been a reduced need for detail at the subclass level. The total number of subclasses has decreased from 285 to 191.

Changes in the structure

SIC (2003) had 17 sections and 62 divisions; SIC (2007) has 21 sections and 88 divisions. At the highest level of SIC some sections can be easily compared with the previous version of the classification. However, the introduction of some new concepts at the section level, for example, the 'Information' section or the grouping of activities linked to environment, makes easy overall comparison between SIC (2007) and its previous version impossible.

The table set out below presents the broad correspondence between the sections of SIC (2003) and SIC (2007) . Please note that this table presents only the rough one-to-one correspondence between the sections: further additional details are necessary to establish the complete correspondence.

	SIC (2003)		SIC (2007)
A B	Agriculture, hunting and forestry Fishing	A	Agriculture, forestry and fishing
C	Mining and quarrying	B	Mining and quarrying
D	Manufacturing	C	Manufacturing
E	Electricity, gas and water supply	D E	Electricity, gas, steam and air conditioning supply Water supply; sewerage, waste management and remediation activities
F	Construction	F	Construction
G	Wholesale and retail trade; repair of motor vehicles, motorcycles and personal and household goods	G	Wholesale and retail trade; repair of motor vehicles and motor cycles
H	Hotels and restaurants	I	Accommodation and food service activities
I	Transport, storage and communications	H J	Transport and storage Information and communication
J	Financial intermediation	K	Financial and insurance activities

K	Real estate, renting and business activities	L M N	Real estate activities Professional, scientific and technical activities Administrative and support service activities
L	Public administration and defence; compulsory social security	O	Public administration and defence; compulsory social security
M	Education	P	Education
N	Health and social work	Q	Human health and social work activities
O	Other community, social and personal services activities	R S	Arts, entertainment and recreation Other service activities
P	Activities of private households as employers and undifferentiated production activities of private households	T	Activities of households as employers; undifferentiated goods- and services-producing activities of households for own use
Q	Extraterritorial organisations and bodies	U	Activities of extraterritorial organisations and bodies

The following table represents the changes, in numerical terms, between SIC (2003) and SIC (2007):

	SIC (2003)	SIC (2007)	Difference
Sections	17	21	+4
Subsections	16	0	−16
Divisions	62	88	+26
Groups	224	272	+48
Classes	514	615	+101
Subclasses	285	191	−94
Manufacturing section			
Sections	1	1	0
Subsections	14	0	−14
Divisions	23	24	+1
Groups	103	95	−8
Classes	242	230	−12
Subclasses	80	51	−29
Other sections			
Sections	16	20	+4
Subsections	2	0	−2
Divisions	39	64	+25
Groups	121	177	+56
Classes	272	385	+113
Subclasses	205	140	−65

The substantial changes between SIC (2003) and SIC (2007) are too numerous to be listed here in their entirety. Nonetheless, the most prominent ones are listed below.

The SIC (2003) sections for agriculture and fishing have been combined. However, the detail under this new section A (Agriculture, forestry and fishing) has been substantially increased. This is in response to continuing requests for more detail in ISIC, mostly due to the fact that agriculture is an important part of the economic structure in many developing countries.

New divisions in manufacturing, representing important new industries or old industries that have increased their economic or social relevance, have been created, such as division 21 (Manufacture of basic pharmaceutical products and pharmaceutical preparations) and division 26 (Manufacture of computer, electronic and optical products). The scope of the latter differs from division 30 (Manufacture of office machinery and computers) in SIC (2003), making it a better tool for statistics on high-tech activities. Other new divisions, such as division 11 (Manufacture of beverages) and 31 (Manufacture of furniture) have resulted from splitting existing divisions and therefore raising their components from group level, as before, to division level.

Most of the remaining divisions in section C (Manufacturing) are unchanged, except SIC (2003) divisions 22 (Publishing, printing and reproduction of recorded media) and 37 (Recycling), of which substantial portions have been moved to other sections (see below).

Repair and installation of machinery and equipment, formerly classified under manufacturing of the corresponding type of equipment, is now identified separately in division 33 (Repair and installation of machinery and equipment). All specialised repair activities are now classifiable separately in SIC, although no high-level aggregate for 'Repair' has been created.

A new section E (Water supply; sewerage, waste management and remediation activities) has been created, which includes the sanitation activities in SIC (2003) division 90, water collection and distribution activities in SIC (2003) division 41, and materials recovery activities, which largely correspond to SIC (2003) division 37. This section now groups activities of common policy interest, but is also based on the actual organisation of these activities in a large number of countries. The detail on these activities has been substantially increased.

The concept of specialised construction activities (also known as special trades) has been introduced in SIC (2007), replacing the division structure of the previous version, which was based largely on the stage of the construction process.

Repair of household goods has been removed from section G (Wholesale and retail trade; repair of motor vehicles and motorcycles) of SIC (2003) to section S (Other service activities) of SIC 2007. However, the exception for classifying trade and repair of motor vehicles and motorcycles in division 45 of SIC (2007) (corresponding to division 50 in SIC (2003)) has been retained for comparability and continuity reasons.

A new section J (Information and communication) has been created, combining activities involving production and distribution of information and cultural products, provision of the means to transmit or distribute these products, as well as data or communications, information technology activities and the processing of data and other information service activities. The main components of this section are publishing activities, including software publishing (division 58), motion picture and sound recording activities (division 59), radio and TV broadcasting and programming activities (division 60), telecommunications activities (division 61) and information technology activities (division 62) and other information service activities (division 63). These activities were included in SIC (2003) sections D (Manufacturing), I (Transport, storage and communications), K (Real estate, renting and business activities) and O (Other community, social and personal service activities), therefore having a strong impact on comparability with the previous SIC version. However, this new treatment of information and communication activities provides a more consistent approach than the previous version of SIC, based on the character of the activities carried out.

In section K (Finance and insurance activities), two classes have been introduced that go beyond the traditional scope of SIC in covering economic production, namely class 64.20 (Activities of holding companies) and 64.30 (Trusts, funds and similar financial entities).

The SIC (2003) section for 'Real estate, renting and business activities' has been split into three sections in SIC (2007) . Real estate is now represented as a stand-alone section (section L) due to its size and importance in the System of National Accounts. The remaining activities have been separated into section M (Professional, scientific and technical activities), covering activities that require a high degree of training and make specialised knowledge and skills available to users, and section N (Administrative and support service activities), covering activities that support general business operations and do not focus on the transfer of specialised knowledge. 'Computer and related activities' (SIC (2003) division 72) are no longer part of this section. Computer repair activities have been grouped with repair of household goods in section S, while software publishing and IT activities have been grouped in the new section J.

The scope of 'Education' (section P) has been changed explicitly to include specialised sport, cultural and other educational services and also specialised support services.

More detail has been added under section Q (Human health and social work activities), creating three divisions instead of one, as in the previous version of SIC. In addition, the focus has been narrowed and includes only human health activities, providing a better tool for measuring this important part of the economy. As a result, veterinary activities have been

removed from this section and put in a division in section M (Professional, scientific and technical activities).

Substantial components of SIC (2003) section O (Other community, social and personal service activities) have been moved to SIC (2007) sections E (Water supply; sewerage, waste management and remediation activities) and J (Information and communication), as described above. The remaining activities have been regrouped in two new sections for 'Arts, entertainment and recreation' (section R) and 'Other service' activities (section S). As a result, activities such as creative arts, library activities and gambling activities have been raised to the division level. Repair of computers and personal and household goods is now included in this new section S.

Correspondence tables: scope and use

Correspondence tables are important tools for comparing statistical data collected and presented using different classifications. They become necessary when the classification changes over time, or when different underlying frameworks do not allow classifications to be closely related. Correspondence tables between different versions of the same classification are used to describe the detailed changes that have taken place in the revision process.

Since SIC is used for the collection and presentation of statistics in many areas, there has been a strong need for correspondence tables between the current SIC and its previous versions. Complete detailed correspondences between SIC (2007) and SIC (2003) , and vice versa, are available in electronic form at: www.statistics.gov.uk/statbase/Product.asp?vlnk=14012

Related classifications

There are other classifications, both national and international, which may be used in conjunction with the industrial classification. These include, for example, the classification of occupations, which relates to the jobs performed by individual workers rather than to the industry in which they work. The workers classified to a particular industry will fall into a number of different categories of an occupational classification and similarly the workers in some occupations may be found in many different industries. The UK uses the *Standard Occupational Classification* (SOC)[3].

A second classification is by sector. In the UK this is the Sector Classification for the National Accounts[4]. The economy is split up into institutional sectors with each economic unit allocated to one of these sectors. The sectors are: General government, Financial corporations, Non-financial corporations (private and public), Non-profit institutions serving households and the 'Rest of the world' sector. In principle, the classification embraces all economic units engaging in transactions in goods and services and financial assets. Thus it includes persons and households and overseas concerns as well as corporations and public bodies. The industrial classification does not make a distinction based on institutional sector but rather brings together units engaged in similar activities, irrespective of ownership. However, the SIC code in conjunction with the legal status and UK/non-UK ownership status of a unit provides an approximation to the sector code.

A third classification is that of products, often a listing of individual products or products grouped according to the industries in which they are principally produced. The number of entries depends on the level of detail required in terms of characteristics such as materials used, quality, size and shape. The amount of detail needed for statistical purposes is much less than would be required for, say, a manufacturer's catalogue. The classification can also be extended to cover not only the production of goods but also economic activities such as distribution, transport and services, although the number of different types of service will normally be much less than the number of different products of the production industries. Each product (good or service) is, in general, classified to only one industry: that in which it is mainly produced. In practice, many units produce not only the goods or services that are principal products of the industries to which the units are classified, but also products mainly produced in other industries.

At the international level, the United Nations has the *Central Product Classification* (CPC)[5]. The main aim of the CPC is to provide a general framework for international comparisons of product statistics. It applies to tradeable and non-tradeable goods and services. The UN documentation of the CPC provides direct links to the *Harmonized Commodity Description and Coding System* (HS) and the *Standard International Trade Classification* (SITC) described below. The European Community, however, preferred a product classification that was closer to the industrial activity classification and devised the *Classification of Products by Activity* (CPA)[6]. The CPA 2008 relates directly to the classification structure in NACE Rev. 2, the first four digits are, with very few exceptions, the same, and has links to CPC via the fifth and sixth digit. The CPA, then, provides details of products by economic activity.

The CPA is linked to the PRODCOM list, which extends the CPA code structure from six to eight digits. PRODCOM is the abbreviation for the EU system of production statistics for mining and manufacturing (that is, excluding services, other than 'industrial services'). The product classification (PRODCOM list), upon which production statistics are based, is drawn up each year by the PRODCOM committee. The headings of the PRODCOM list are derived from the Combined Nomenclature

(CN), but their code is a further breakdown of the CPA code. PRODCOM headings are coded using an eight digit numerical code, the first six digits of which are identical to those of the CPA code. The PRODCOM list is therefore linked to, and consistent with, CPA. The link with CPA emphasises the link with NACE, enabling the enterprises producing the products to be identified, while the link with CN allows comparisons between production statistics and foreign trade statistics[7].

In addition to the product lists associated with industrial activity classifications, there are also classifications recording imports and exports. The United Nations *Standard International Trade Classification* (SITC) and the *Harmonized Commodity Description and Coding System* (HS) with which it is correlated have been widely adopted throughout the world as the basis for national classifications both for tariff and trade statistics purposes. The HS was devised by the World Customs Organisation to support international trade and to meet the requirements of customs authorities, statisticians, carriers and producers. The SITC is the most commonly published format for foreign trade statistics. The member states of the European Community used the more detailed Combined Nomenclature (CN) to collect trade data through their customs procedures. The CN was developed directly from the HS and was used for the collection of both intra-Community and external trade statistics.

Use of the UK SIC (2007)

SIC classifications are determined according to the principal activity of a unit. Ideally, this is based on value added; in practice a proxy, such as turnover or employment, is frequently needed. Detailed guidance is set out in the section, 'Rules for Classifying Statistical Units'.

UK SIC (2007) Indexes

As with previous versions of the SIC, there is a separate publication containing detailed lists of activities and, in some cases, the products of those activities, contained in UK SIC (2007) at: www.statistics.gov.uk/statbase/Product.asp?vlnk=14012

References

1. NACE Rev. 1 Regulation No 3037/1990 was published in the Official Journal of the European Communities L 293 Volume 33, 24 October 1990 (ISSN 0378-6978). In January 2003, a minor revision, NACE Rev. 1.1, was published in the Official Journal of the European Communities L6/3 10 January 2002

2. NACE Rev. 2 Regulation (EC) No 1893/2006 was published in the Official Journal of the European Communities L 393 Volume 49, 30 December 2006 (ISSN 1725-2555).

3. The current *Standard Occupational Classification* (SOC) was published in 2000 (Volume 1; ISBN 0-11-621388-4).

4. The *UK Sector Classification for the National Accounts* is in Sector classifications Guide MA 23.

5. A revised version of the CPC (*Central Product Classification*), CPC ver.2, was adopted by the United Nations Statistical Commission in March 2006.

6. The *Classification of Products by Activity* (CPA) Regulation (EEC) No 3696/93 was published in the Official Journal of the European Communities L342 Volume 36, 31 December 1993. A proposal for a regulation of the European Parliament and of the Council establishing a new statistical classification of products by activity (CPA 2008) and repealing Council Regulation (EEC) No. 3696/93 was submitted to the Council of the European Parliament on 3 November 2006.

7. The PRODCOM Regulation (EEC) No 3924/91 was published in the Official Journal of the European Communities L 374/1, 31 December 1991

Rules for classifying statistical units

Contents	Paragraphs
Classification: definitions and rules	

Classification: definitions principals and rules

General

Principles used in constructing NACE and followed in SIC (2007)

(Note: The UK SIC (2007) is based exactly on NACE Rev. 2 down to and including the four digit class level. Where it was thought necessary or helpful, a fifth digit has been added to the SIC to form subclasses of the NACE four digit classes. Throughout the following rules, therefore, all references to NACE categories should be taken to apply equally to the equivalent SIC categories and to the SIC five digit subclasses.)

Criteria adopted developing NACE

1. The criteria used to define and delineate classification categories at any level depend on many factors, such as potential use of the classification and availability of data. These criteria are applied differently at different levels of the classification: the criteria for detailed levels of the aggregation consider similarities in the actual production process, while this is largely irrelevant at more aggregated levels of the classification.

Criteria for classes

2. The criteria concerning the manner in which activities are combined in, and allocated among, production units are central in the definition of classes (most detailed categories) of NACE. They are intended to ensure that the classes of NACE will be relevant for the detailed industrial classification of units, and that the units falling into each class will be as similar, in respect of the activities in which they engage, as is feasible.

3. NACE Rev. 2, reflecting the fourth revision of ISIC, generally gives more importance to the production process in the definition of individual classes. This means that activities are grouped together when they share a common process in producing goods or services, using similar technologies.

4. In addition, the classes of NACE are defined so that the following two conditions are fulfilled whenever possible:

    a. The production of the category of goods and services that characterises a given class accounts for the bulk of the output of the units classified according to that class.

    b. The class contains the units that produce most of the category of goods and services that characterise it.

5. Another major consideration in defining classes in NACE is the relative importance of the activities to be included. In general, separate classes are provided for activities that are prevalent in most EU countries, or that are of particular importance in the world economy. To attain international comparability, certain classes have been introduced in the structures of ISIC and, therefore, included in NACE.

Criteria for groups and divisions

6. Unlike for classes, the actual production process and technology used in production activities become less important as a criterion for grouping them at more aggregated levels. At the highest level (sections), the general characteristics of the goods and services produced, as well as the potential use of the statistics, for instance in the framework of SNA and ESA, becomes an important factor.

7. The main criteria applied in delineating groups and divisions of NACE concern the following characteristics of the activities of production units:

    • the character of the goods and services produced
    • the uses to which the goods and services are put, and
    • the inputs, the process and the technology of production

8. In the case of the character of the goods and services produced, account is taken of the physical composition and stage of fabrication of the items and the needs served by them. Distinguishing categories of NACE in terms of the nature of goods and services produced provides the basis for grouping production units according to similarities in, and links between, the raw materials consumed and the sources of demand and markets for the items.

9. The weight assigned to the criteria described above varies from one category to another. In a number of instances (for example, food manufacturing, the textile, clothing and leather industries; machinery and equipment manufacturing; as well as the service industries) the three specific aspects are so closely related that the problem of assigning weights to the criteria does not arise. In the case of intermediate products, the physical compositions, as well as the stage of fabrication of the items, are often given the greatest weight. In the case of goods with complicated production processes, the end use, the technology and the organization of production of the items are frequently given priority over the physical composition of the goods.

Economic activity

10. An economic activity takes place when resources such as capital goods, labour, manufacturing techniques or intermediary products are combined to produce specific goods or services. Thus, an economic activity is characterised by an input of resources, a production process and an output of products (goods or services).

11. An activity as defined here may consist of one simple process (for example, weaving), but may also cover a whole range of sub-processes, each mentioned in different categories of the classification (for example, the manufacturing of a car consists of specific activities such as casting, forging, welding, assembling, painting, etc.). If the production process is organised as an integrated series of elementary activities within the same statistical unit, the whole combination is regarded as one activity.

12. NACE does not *per se* provide categories for specific types of statistical units: units may perform several economic activities, and can be defined in different ways according to specific characteristics (related for example, to location, see the section on 'statistical units' below).

Principal, secondary and ancillary activities

13. A unit may perform one or more economic activities described in one or more categories of NACE.

14. The principal activity of a statistical unit is the activity which contributes most to the total value added of that unit. The principal activity is identified according to the top-down method (see paragraph 40) and does not necessarily account for 50 per cent or more of the unit's total value added.

15. A secondary activity is any other activity of the unit that outputs goods or services suitable for delivery to third parties. The value added of a secondary activity must be less than that of the principal activity.

16. A distinction should be made between principal and secondary activities, on the one hand, and ancillary activities, on the other. Principal and secondary activities are generally carried out with the support of a number of ancillary activities, such as accounting, transportation, storage, purchasing, sales promotion, repair and maintenance, etc. Thus, ancillary activities are those that exist solely to support the principal or secondary economic activities of a unit, by providing goods or services for the use of that unit only.

17. An activity is ancillary if it fulfils all of the following conditions:

    a. it serves only the unit or units referred to

9

b. the inputs contribute to the costs of the unit

c. the outputs (usually services, seldom goods) are not part of the unit's final product and do not generate gross fixed capital formation

d. a comparable activity on a similar scale is performed in similar production units

18. For instance the following are not to be regarded as ancillary activities:

a. production of goods and services that are part of capital formation; for example, construction work for own account, which would be separately classified to construction if data were available, and software production

b. production of outputs, a significant part of which is sold on the market, even if part of it is consumed in connection with principal activities

c. production of goods or services which subsequently become an integral part of the output of the principal or secondary activity (for example, production of boxes by a department of an enterprise for packing its products)

d. production of energy (an integrated power station or coking plant), even if the whole output is consumed by the parent unit

e. purchase of goods for resale in an unaltered state

f. research and development, as these activities do not provide a service that is consumed in the course of current production

19. In all these cases, where separate data are available, separate units should be distinguished and recognised as kind-of-activity units (see next section), and then classified according to their activity.

Kind of ownership 20. NACE does not draw distinctions according to the kind of ownership of a production unit or its type of legal organisation or mode of operation, because such criteria do not relate to the characteristics of the activity itself. Units engaged in the same kind of economic activity are classified in the same category of NACE, irrespective of whether they are (part of) incorporated enterprises, individual proprietors or government, whether or not the parent enterprise is a foreign entity and whether or not the unit consists of more than one establishment. Therefore, there is no link between NACE and the Classification of Institutional Units in the System of National Accounts (SNA) or in the European System of Accounts (ESA). However, the SIC code in conjunction with the legal status and UK/non-UK ownership status of a unit provides an approximation to the sector code.

21. The manufacturing activities are described independently of whether the work is performed by power-driven machinery or by hand, or whether it is done in a factory or in a household. Modern versus traditional is not a criterion for NACE.

22. NACE does not distinguish between formal and informal or between legal and illegal production. Classifications according to kind of legal ownership, kind of organisation or mode of operation may be constructed independently. Cross-classification with NACE could provide useful extra information.

Market and non-market 23. In general, NACE does not differentiate between market and non-market activities, as defined in the SNA/ESA, even if this distinction is an important feature of the SNA/ESA. A breakdown of economic activities according to this principle is useful in any case where data are collected for activities that take place on both a market and a non-market basis. This criterion should then be cross-classified with the categories of NACE. Non-market services in NACE are only provided by government organisations or non-profit institutions serving households, mostly in the field of education, health, social work, etc.

24. NACE includes categories for the undifferentiated production of goods and services by households for their own use. These categories may refer, however, to only a portion of households' economic activities, as clearly identifiable household activities are classified in other parts of NACE.

Statistical units

Definitions of statistical units

25. In order to draw a complete statistical picture of the economy, a wide range of information is required, and the organizational level at which it is feasible to collect the information varies according to the type of data. For example, profits data for a company may be available from only one geographically central location referring to several different locations, whereas product sales data may be available referring to each of the separate locations. To observe and analyse the data satisfactorily, it is therefore necessary to define a system of statistical units. These form the reference building blocks in respect of which data can be collected and classified according to NACE.

26. Different types of statistical units meet different needs, but each unit is a specific entity, which is defined in such a way that it can be recognised and identified and not confused with any other unit. It may be an identifiable legal or physical entity or, as for example in the case of the unit of homogeneous production, a statistical construct.

27. The following are the statistical units that are described in the Council Regulation on statistical units[1]:

   a. the enterprise group

   b. the enterprise

   c. the kind-of-activity unit (KAU)

   d. the local unit

   e. the local kind-of-activity unit (local KAU)

   f. the institutional unit

   g. the unit of homogeneous production (UHP)

   h. the local unit of homogeneous production (local UHP)

The relationship between the different types of statistical units is illustrated in the following table:

	One or more locations	**A single location**
One or more activities	Enterprise Institutional unit	Local unit
One single activity	KAU	Local KAU
	UHP	Local UHP

---

1 Council Regulation (EEC) No 696/93 of 15 March 1993 on the statistical units for the observation and analysis of the production system in the Community (OJ No L 76, 30.3.1993, p 1).

The system of administrative and statistical units can be illustrated as follows:

Classification rules for activities and units

Classification rules for activities and units

Basic classification rules   28.   One NACE code is associated to each unit recorded in statistical business registers[2], according to its principal economic activity. The principal activity is the activity which contributes most to the value added of the unit. The assignment of the NACE code is helped by: the NACE explanatory notes, decisions taken by the NACE management committee, correspondence tables and reference to other classification systems such as the ISIC, CPA, HS, CN, etc.

29.   In the simple case where a unit performs only one economic activity, the principal activity of that unit is determined by the category of NACE which describes that activity. If the unit performs several economic activities (other than ancillary activities, see paragraphs 10 to 17), the principal activity is determined on the basis of the value added associated to each activity, according to the rules presented below.

30.   Value added is the basic concept for the determination of the classification of a unit according to economic activities. The gross value added is defined as the difference between output and intermediate consumption. Value added is an additive measure of the contribution of each economic unit to the Gross Domestic Product (GDP).

Value added substitutes   31.   In order to determine the principal activity of a unit, the activities carried out by the unit and the corresponding share of value added have to be known. Sometimes it is not possible to obtain the information on value added associated with the different activities carried out, and the determination of the activity classification has to be done by using substitute criteria. Such criteria could be:

a.   Substitutes based on output:

2. Council Regulation (EEC) No 2186/93.

- gross output of the unit that is attributable to the goods or services associated with each activity
- value of sales or turnover of those groups of products falling within each activity

b. Substitutes based on input:

- wages and salaries attributable to the different activities (or income of self-employed)
- number of staff involved in the different economic activities of the unit
- time worked by staff attributable to the different activities of the unit

32. Such substitute criteria should be used as proxies for the unknown value added data, to obtain the best approximation possible compared to the result which would have been obtained on the basis of the value added data. The use of substitute criteria does not change the methods used to determine the principal activity, as they are only operational approximations of value added data.

33. However, the simple use of the above listed substitute criteria may be misleading. This will always be the case when the structure of the substitute criteria is not directly proportional to the (unknown) value added.

34. When using sales (turnover) as a proxy for value added, it should be taken into account that in certain cases turnover and value added are not proportional. For example, turnover in trade usually has a much lower share of value added than turnover in manufacturing. Even within manufacturing the relation between sales and the resulting value added may vary between and within activities. For some activities turnover is defined in a specific way which prevents useful comparison with other activities, for example, financial intermediation activities or insurance activities. The same considerations should be borne in mind when using gross output data as substitute criteria.

35. Many units perform trade and other activities. In such cases trade turnover figures are the most unsuitable indicators for the unknown value added share of the trade activity. A much better indicator is the gross margin (difference between the trade turnover and purchases of goods for resale adjusted by changes in stocks). However, the trade margins may vary within a single wholesale and retail trade and also between trade activities. In addition, consideration must be given to the specific classification rules for trade as set out in paragraphs 103 to109.

36. Similar precautions have to be considered when input-based substitute criteria are applied. The proportionality between wages and salaries or employment on one hand, and value added on the other is not reliable if the labour intensity of the various activities is different. Labour intensity may vary substantially between different economic activities and also between activities of the same NACE class. Example: the production of a good by hand vs. the production of a good using a mechanised process.

Multiple and integrated activities

37. Instances may arise where considerable proportions of the activities of a unit are included in more than one class of NACE. These cases may result from the vertical integration of activities (for example, tree felling combined with sawmilling, or activities of a clay pit combined with brickworks), or the horizontal integration of activities (for example, manufacture of bakery products combined with manufacture of chocolate confectionery), or any combination of activities within a statistical unit. In these situations, the unit should be classified according to the rules set out in this section.

38. If a unit performs activities falling in only two different NACE categories, there will always be one activity that accounts for more than 50 per cent of value added, except in the highly unlikely case that both of the activities in question have equal shares of 50 per cent.

The activity that represents more than 50 per cent of the value added is the principal activity and determines the NACE Rev. 2 classification of the unit.

39. In the complex case where a unit performs more than two activities falling into more than two different positions of NACE, with none of them accounting for more than 50 per cent of value added, the activity classification of that unit has to be determined by using the 'top-down' method, as described below.

The top-down method　40. The top-down method follows a hierarchical principle: the classification of a unit at the lowest level of the classification must be consistent with the classification of the unit at the higher levels of the structure. To satisfy this condition the process starts with the identification of the relevant position at the highest level and progresses down through the levels of the classification in the following way:

a. Identify the section which has the highest share of the value added.

b. Within this section identify the division which has the highest share of the value added.

c. Within this division identify the group which has the highest share of the valued added.

d. Within this group identify the class which has the highest share of value added.

Example: a unit carries out the following activities (shares in terms of value added):

Section	Division	Group	Class	Description of the class	% share
C	25	25.9	25.91	Manufacture of steel drums and similar containers	10
	28	28.1	28.11	Manufacture of engines and turbines, except aircraft, vehicle and cycle engines	6
		28.2	28.24	Manufacture of power-driven hand tools	5
		28.9	28.93	Manufacture of machinery for food, beverage and tobacco processing	23
			28.95	Manufacture of machinery for paper and paperboard production	8
G	46	46.1	46.14	Agents involved in the sale of machinery, industrial equipment, ships and aircraft	7
		46.6	46.61	Wholesale of agricultural machinery, equipment and supplies	28
M	71	71.1	71.12	Engineering activities and related technical consultancy	13

Identify the main section among

Section C	Manufacturing	(52%)
Section G	Wholesale and retail trade; repair of motor vehicles and motorcycles	(35%)
Section M	Professional, scientific and technical activities	(13%)

Identify the main division within main section C:

Division 25	Manufacture of fabricated metal products, except machinery and equipment	10%
Division 28	Manufacture of machinery and equipment n.e.c[3].	42%

Identify the main group within the main division 28:

Group 28.1	Manufacture of general-purpose machinery	6%
Group 28.2	Manufacture of other general-purpose machinery	5%
Group 28.9	Manufacture of other special-purpose machinery	31%

3　Not elsewhere classified.

Identify the main class within the main group 28.9:

Class 28.93   Manufacture of machinery for food, beverage and tobacco processing   23%

Class 28.95   Manufacture of machinery for paper and paperboard production   8%

Therefore the correct class is 28.93 'Manufacture of machinery for food, beverage and tobacco processing', although the class with the biggest share of value added is class: 46.61 'Wholesale of agricultural machinery, equipment and supplies'.

The following diagram represents the decision path followed in the example.

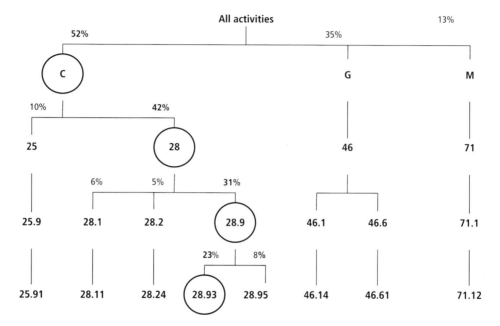

41. Because of the difficulties between ISIC and NACE at group and class levels, the application of the top-down method to NACE may give allocations different from those which would be obtained applying the method to ISIC. If possible, the method should be followed first in terms of ISIC, identifying the ISIC class, and then in terms of NACE. This would ensure alignment to world classifications.

42. When applied to section G 'Wholesale and retail trade', a specific adaptation of the top-down method is required. See paragraphs 103 to 109 below for details.

Changes of the principal activity of the unit

43. Units may change their principal activity, either at once or gradually over a period of time, either because of seasonal factors or because of a management decision to vary the pattern of output. While all these cases call for the classification of the unit to be changed, too frequent changes could result in inconsistencies between short term (monthly and quarterly) and longer term statistics, making their interpretation extremely difficult.

44. Whenever a unit performs two activities both contributing to around 50 per cent of the value added, a stability rule has been established in order to avoid frequent changes not reflecting a substantial change in the economic reality. According to this rule, the change of the principal activity should be made when the current one had been accounting for less than 50 per cent of the value added for at least two years.

Treatment of vertically integrated activities

45. Vertical integration of activities occurs where the different stages of production are carried out in succession by the same unit and where the output of one process serves as input to the next. Common examples of vertical integration include tree felling and subsequent sawmilling, a clay pit combined with brickworks, or production of wearing apparels in a textile mill.

46. When applying NACE Rev. 2, vertical integration should be treated like any other form of multiple activities, that is, the principal activity of the unit is the activity accounting for the largest share of value added, as determined according to the top-down method. This treatment has changed from previous versions of NACE. For vertical integration of specific situations in Agriculture, see paragraph 102.

47. If value added or substitutes cannot be determined for the individual steps in a vertically integrated process directly from accounts compiled by the unit itself, comparisons with similar units could be used. Alternatively, valuation of intermediate or final products could be based on market prices.

Treatment of horizontally integrated activities

48. Horizontal integration of activities occurs when activities are carried out simultaneously using the same factors of production. The principle of value added, has to be applied, following the top-down method, and the same precautions for using substitutes as listed above apply here.

Activity specific rules, definitions and rules for classification of units outsourcing their production

Terminology Outsourcing

49. Contractual agreement according to which the principal requires the contractor to carry out specific tasks, such as parts of a production process or even the full production process, employment services or support functions.

50. The term outsourcing is also valid if the contractor is a subsidiary unit and whether or not the tasks are carried out on market conditions.

51. The *principal* and the *contractor* may be located within the same economic territory or in different economic territories. The actual location does not affect the classification of either one of these unit.

Principal

52. A unit that enters into a contractual relationship with another unit (*contractor*) for that other unit to carry out specific tasks, such as parts of a production process or even the full production process, employment services or support functions.

Contractor

53. A unit that carries out specific tasks, such as parts of a production process or even the full production process, employment services or support functions on a contractual relationship with a *principal*. The term *subcontractor* is also used. In NACE the activities performed by the *contractor* are denominated 'on a fee or contract basis'.

Manufacturing activities

54. The physical and/or chemical transformation of materials, substances or components into new products. The material, substances or components are raw materials which are products of agriculture, forestry, fishing or mining as well as products and semi-finished products of other manufacturing activities.

Subcontractor

55. See *contractor* as defined in paragraph 53.

Classification rules

56. These rules set out how to classify the outsourcing activities of principals and contractors, as defined in paragraphs 52 and 53. It is important to note that the rules apply only to the classification of the outsourcing activities and that where principals or contractors are also involved in other activities, their overall activity classification must be determined by applying the value added guidance set out in paragraphs 30 to 48 to all of their activities.

Outsourcing of parts of a manufacturing production process

57. A principal delegates a part of a manufacturing production process to a contractor.

Codification rules

58. The principal has to be classified as if he were carrying out the complete production process.

59. The contractor is classified with units producing the same goods or services for their own account.

Outsourcing of the complete manufacturing production process (i)

60. A principal who owns the main material inputs subcontracts the complete manufacturing production process of products to another unit.

Codification rules

61. The principal who owns the main material inputs (for example, textiles and buttons for the production of apparel, wood and metal accessories for the manufacture of furniture) and thereby owns the final outputs, but who has done the production by contractors, is classified in NACE section C (Manufacturing), specifically to the class that corresponds the full production process.

62. The contractor is classified with units producing the same goods for their own account.

Outsourcing of the complete manufacturing production process (ii)

63. A principal who doesn't own the material inputs subcontracts the complete manufacturing of products to another unit.

Codification rules

64. The principal that has the production done by others and who does not own the material inputs, should be classified in section G 'Wholesale and trade' (depending on the activity and the specific good sold). The contractor is classified with units producing the same goods for their own account.

Outsourcing of construction activities

65. A principal subcontracts the construction work to other units, but remains overall responsible for the construction process.

Codification rules

66. Both the principal and the contractors are classified in NACE section F (Construction), specifically to the class that corresponds to the construction activities carried out.

Outsourcing of support functions

67. A principal carries out the whole or a part of the production process (of a good or a service) but delegates certain support functions, such as accounting or computer services, to a contractor. These support functions are not part of the core production process, they do not directly lead to the final good or service, but support the general functioning of the principal as a production unit.

Codification rules

68. The principal is classified to the same NACE code that represents the core production process. The contractor is classified to the specific activity he is carrying out, for example, NACE 69.20 (Accounting, bookkeeping and auditing activities; tax consultancy), NACE 62.02 (Computer consultancy activities) etc.

Outsourcing of employment services

Codification rules

69. In the case of outsourcing of employment services a distinction should be drawn between outsourcing on a temporary or on a long-term and permanent basis.

    a. In outsourcing on a temporary basis, the principal is classified on the basis of the activity actually performed (for example, manufacturing).

    The contractor is classified under NACE 78.20 (Temporary employment agency activities).

    b. In outsourcing on a long-tern or permanent basis, the principal is classified on the basis of the activity actually performed (for example, manufacturing).

    The contractor is classified under NACE 78.30 (Other human resources provision).

Outsourcing of service-producing activities

70. The principal subcontracts a part or the complete provision of services (excepted support functions, see rule 'Outsourcing of support functions' above) to another unit.

Codification rules

71.

    a. The principal who outsources a part of the service-producing activities has to be classified as if he were performing the whole service process. The contractor is classified according to the portion of the services provision he is undertaking.

    b. If the principal subcontracts the whole services activity, both the principal and the contractor are classified as if they were carrying out the complete services activity.

Outsourcing of the complete agricultural or animal production process (i)

72. A principal who owns the seeds or the plants (seedlings, cuttings, sprays) or the fruit trees (including grapevines) or herds of animals subcontracts the complete agricultural or animal production process of agricultural products to another unit.

Codification rules

73. The principal who owns the seeds or the plants (seedlings, cuttings, sprays) or the fruit trees (including grapevines) or herds of animals and thereby owns the final outputs, but who has the production done by contractors, is classified in division 01, specifically to the class that covers the full production process.

74. The contractor is classified in the appropriate class of group 01.6.

Outsourcing of the complete agricultural or animal production process (ii)

75. A principal who doesn't own the seeds or the plants (seedlings, cuttings, sprays) or the fruit trees (including grapevines) or herds of animals subcontracts the complete production process to another unit.

Codification rules

76. The principal who has the production done by others and who does not own the seeds or the plants (seedlings, cuttings, sprays) or the fruit trees (including grapevines) or herds of animals, is classified in section G 'Wholesale and trade' (depending on the activity and the specific good sold.

77. The contractor is classified with units producing the same goods for their own account.

Outsourcing of the complete forestry production process (i)

78. A principal who owns the trees subcontracts the complete forestry production process of forest products to another unit.

Codification rules

79. The principal who owns the trees and thereby owns the final outputs, but who has the production done by contractors, is classified in division 02, specifically to the class that covers the full production process.

80. The contractor is classified in class 02.40.

Outsourcing of the complete forestry production process (ii)

81. A principal who doesn't own the trees sub-contracts the complete production process to another unit.

Codification rules

82. The principal who has the production done by others and who does not own the trees is classified in section G 'Wholesale and trade' (depending on the activity and the specific good sold).

83. The contractor is classified with units producing the same goods for their own account.

Outsourcing of the complete aquaculture production process (i)

84. A principal who owns the fry subcontracts the complete aquaculture production process of aquaculture products to another unit.

Codification rules

85. The principal who owns the fry and thereby owns the final outputs, but who has the production done by contractors, is classified in group 03.2, specifically to the class that covers the full production process.

86. The contractor is classified in the appropriate class of group 03.2.

Outsourcing of the complete aquaculture production process (ii)

87. A principal who doesn't own the fry subcontracts the complete production process to another unit.

Codification rules

88. The principal who has the production done by others and who does not own the fry is classified in section G 'Wholesale and trade' (depending on the activity and the specific good sold).

89. The contractor is classified with units producing the same goods for their own account.

Outsourcing of the complete energy production process (i)

90. A principal who owns the energy sources material (such as oil, coal, gas, wood, agricultural residual, etc.) subcontracts the complete energy production process of energy products to another unit.

Codification rules

91. The principal who owns the energy sources material (such as oil, coal, gas, wood, agricultural residual, etc.) and thereby owns the final outputs, but who has the production done by contractors, is classified in section D, specifically to the class that corresponds the full production process.

92. The contractor is classified in the appropriate class of section D.

Outsourcing of the complete energy production process (ii)

93. A principal who doesn't own the energy sources material (such as oil, coal, gas, wood, agricultural residual, etc.) subcontracts the complete production process to another unit.

Codification rules

94. The principal who has the production done by others and who does not own the energy sources material (such as oil, coal, gas, wood, agricultural residual, etc.) is classified in section D class 35.14 or 35.23 (depending on the activity and the specific good sold).

95. The contractor is classified with units producing the same goods for their own account.

Outsourcing of other complete goods-production process

96. In the cases of fishing, mining and quarrying and water supply, the principal who has the production done by others is classified in section G 'Wholesale and trade' (depending on the activity and the specific good sold).

97. The contractor is classified with units producing the same goods or services for their own account.

On-site installation

98. Units principally engaged in the installation or assembly of items or equipment in buildings for their functioning are classified in the construction section (division 43).

99. Installation of machinery and other equipment other than those linked to the functioning of buildings (or civil engineering works) is classified in group 33.2 'Installation of industrial machinery and equipment'.

Repair and maintenance

100. Units that repair or maintain goods are classified into one of the following categories, depending on the types of goods:

group 33.1 Repair of fabricated metal products, machinery and equipment

division 43 Specialised construction activities

group 45.2 Maintenance and repair of motor vehicles

division 95 Repair of computers and personal and household goods

101. Units that overhaul aircrafts, locomotives and ships are classified in the same class as the units that manufacture them.

Section specific rules and definitions

102. This section presents rules and definitions to be taken into account when classifying units in specific sections. General descriptions, definitions and characteristics of sections are presented in the corresponding explanatory notes.

Section A: Agriculture, Forestry and Fishing

103. In 'Agriculture', one frequent situation where the breakdown of the value added presents difficulties is when the unit produces grapes and manufactures wine from the self-produced grapes, or when it produces olives and manufactures oil from the self-produced olives. In these cases the most suitable proxy variable is the 'number of hours worked', and its application to these vertically integrated activities would generally lead to the classification of the units in agriculture. In the same case for other agricultural products, units will be classified in agriculture by convention, in order to guarantee a harmonised treatment.

Section G: Wholesale and retail trade; repair of motor vehicles and motorcycles

104. In section G, trade is distinguished between wholesale and retail sale, apart from the trade in motor vehicles. It may happen that a unit performs horizontally integrated trade activities under various possible forms: both wholesale and retail sale, or sale in store and not in stores, or a wide variety of goods are sold. If the goods sold by the unit do not comprise a unique class accounting for at least 50 per cent of the value added, then the application of the top-down method requires special caution and the consideration of additional levels.

105. Within division 46 'Wholesale trade', first an additional level of distinction has to be considered: group 46.1 'Wholesale on a fee or contract basis' and the aggregation of groups 46.2-46.9. Therefore, the first decision to take is the allocation of the unit to one of these two possibilities, on the basis of the value added principle. If the choice falls on aggregation level 46.2-46.9, then the second step consists in deciding between 'non-specialised' vs. 'specialised' (see below). Finally, the choice has to be made, always applying the top-down method, among groups and classes.

106. The diagram below represents the decision tree to be used for the allocation of a unit to a specific class in division 46 'Wholesale trade':

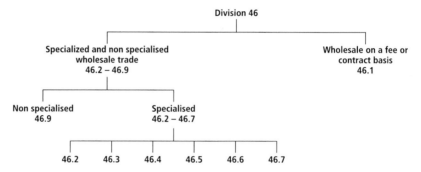

Further subdivided according to the correspondent products

107. Within division 47 'Retail trade', first an additional level of distinction has to be considered: the aggregation of groups 47.1 to 47.7 'Retail trade in stores' and the aggregation of groups 47.8-47.9 'Retail sale not in stores'. Therefore, the first decision to take is the allocation of the unit to one of these two possibilities on the basis of the value added principle. If the choice falls on the aggregation level 'Retail sale in stores', then the second step consists of deciding between 'non-specialised' and 'specialised' (see below). Finally, the choice has to be made, always applying the top-down method, to the groups and classes.

108. The diagram below represents the decision trees to be used for the allocation of a unit to a specific class in division 47 'Retail trade'.

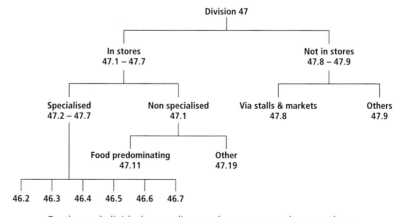

Further subdivided according to the correspondent products

21

109. Both in wholesale and in retail sale trade, the distinction between 'specialised' and 'non-specialised' is based on the number of classes comprising the goods sold, where the classes to be considered each account for at least 5 per cent (and less than 50 per cent) of the value added:

    a. If the products sold comprise up to four classes in any of the groups 46.2 to 46.7 (for wholesale) or 47.2 to 47.7 (for retail sale), the unit is considered to be in 'specialised trade'. It is then necessary to determine the principal activity applying the top-down method on the basis of the value added, selecting first the main group and then the class within that group:

Class	Case A	Case B	Case C
47.21	30%	30%	20%
47.25	5%	15%	5%
47.62	45%	40%	35%
47.75	20%	15%	40%
Final allocation	Class 47.62	Class 47.21	Class 47.75

    b. If the products sold comprise five or more classes in any of the groups 46.2 to 46.7 (for wholesale) or 47.2 to 47.7 (for retail sale), then the unit should be classified as 'non-specialised'. In retail trade, it is therefore allocated to group 47.1. If food, beverages and tobacco represent at least 35 per cent of value added, allocation will be made to NACE Rev. 2 class 47.11. In all other cases allocation should be to class 47.19.

Class	Case A	Case B	Case C
47.21	5%	20%	5%
47.22	10%	15%	5%
47.42	15%	10%	45%
47.43	25%	10%	40%
47.54	45%	45%	5%
Final allocation	Class 47.19	Class 47.11	Class 47.19

110. The allocation rules are always based on the retail activity of the unit. If, in addition to its retail trade, a unit has a secondary activity, the allocation of the unit to the appropriate class is determined only by the conditional composition of its retail activity.

Sections K: Financial and insurance activities, and M: Professional, scientific and technical activities

111. In section K (Finance and insurance activities), two classes have been introduced that go beyond the traditional scope of NACE in covering economic production, namely class 64.20 (Activities of holding companies) and 64.30 (Trusts, funds and similar financial entities). Units classified in these two classes do not have any revenue from the sale of products, and usually do not employ staff (except possibly one or a few persons acting as legal representatives). Sometimes these units are called 'brass plates', or 'postal boxes' or 'empty boxes', or 'special purpose entities – SPE', as they just have a name and an address. They are numerous in some countries because of tax advantages.

112. When classifying a unit according to these two classes, attention should be paid also to other classes (two of them in section M, division 70) namely classes 70.10 'Activities of head offices' and 70.22 'Business and other management consultancy activities'.

113. More specifically:

class 64.20 'Activities of holding companies' refers to activities of holding companies, whose principal activity is owning the group, and that do not administer or manage the group

class 64.30 'Trusts, funds and similar financial entities' is very particular in NACE, as it does not refer to an economic activity, but to units

class 66.30 'Fund management activities' includes activities carried out on a fee or contract basis

class 70.10 'Activities of head offices' includes the overseeing and managing of the related units, exercising operational control and day-to-day managing

class 70.22 'Business and other management consultancy activities' includes consultancy activities related to issues like corporate strategic and organisational planning, marketing objectives and policies, human resources policies etc

114. The principal activity of a unit performing several activities among those just mentioned should be made, as usual, on the basis of the value added principle. It should be noted that capital gains do not constitute value added, and therefore they should not be considered. The introduction of the above mentioned classes represents a major change from NACE 1.1.

Section O: Public administration and defence; compulsory social security

115. NACE does not make any distinction regarding the institutional sector (as defined in SNA and ESA) in which the institutional unit is classified. Moreover, there is no NACE category that describes all activities carried out by the government as such. Consequently, not all government bodies are automatically classified in section O (Public administration and defence; compulsory social security). Units carrying out activities at the national, regional or local levels that are specifically attributable to other areas of NACE are classified in the appropriate section. For example, a school for secondary school administered by central or local government is allocated to group 85.3 (section P), and a public hospital is allocated to class 86.10 (section Q). On the other hand, section O is not restricted to government bodies: private units performing typical 'public administration activities' are also classified here.

Section T: Activities of households as employers; Undifferentiated goods- and services-producing activities of households for own use

116. Division 97 includes only the activities of private households as employers of domestic personnel. The output of this activity is considered as production in the SNA, and for this purpose and for certain surveys this division has been included in NACE Rev. 2. The activities of domestic personnel are not classified here: for instance, baby-sitting activities are to be classified in 88.91, washing of textile in 96.01, and valet activities in 96.09 etc.

117. The need to describe activities for own use has emerged in data collections such as labour force or time use surveys. While market activities should generally be described according to existing rules for identifying the correct NACE code, the application of these rules to activities for own use has proved difficult because, in contrast to market activities, it is difficult to quantify the value added. These activities often combine agricultural, construction, textile manufacturing, repair and other services. Division 98 (Undifferentiated goods- and services-producing activities of private households for own use) corresponds

to divisions 96 and 97 of NACE Rev. 1, which where introduced in order to cover these activities. Division 98 is not relevant in EU business statistics, but is relevant in data collections covering household and subsistence activities.

Glossary

118. This glossary gives a further description of some of the terms used throughout the NACE Rev. 2 Introduction and Explanatory Notes. Every attempt has been made to ensure that the descriptions are consistent with the definitions of the terms when used elsewhere, but these descriptions are not intended to give all-purpose definitive meanings of the words. The purpose of this glossary is merely to help the user of NACE to interpret it correctly.

119. **By-product:** An exclusive by-product is a product technologically linked to the production of other products in the same group, but which is not produced in any other group (for example, molasses linked to the production of sugar). Exclusive by-products are used as inputs for the manufacture of other products. An ordinary by-product (that is, a by-product which is not exclusive to a single group) is a product technologically linked to the production of other products, but which is produced in several groups (for example, the hydrogen produced during petroleum refining is technologically linked to that produced in petrochemical manufacture and coal carbonisation and identical to that produced in the group comprising other basic chemical products).

120. **Commodity:** A commodity is a transportable good that may be exchanged. It may be one of a run from a production line, a unique item (Mona Lisa) or the material medium for a service (software diskette). This is the concept used for customs classifications.

121. **Capital goods**: Capital goods are goods, other than material inputs and fuel, used for the production of other goods and/or services. They include factory buildings, machinery, locomotives, lorries and tractors. Land is not usually regarded as a capital good.

122. **Industrial process**: A transformation process (whether physical, chemical, manual or whatever) used in the manufacture of new products (whether consumer, intermediate or investment goods), in the processing of used products or in the provision of services to industry as defined in sections B (Extractive industries), C (Manufacturing industry), D (Production and distribution of electricity, gas and steam) E (Water supply; sewerage, waste management and remediation activities) and F (Construction industry).

123. **Machinery: domestic or household**: Machinery and equipment of a type designed principally for use by private households, for example, household washing machines.

124. **Machinery: industrial**: Machinery and equipment of a type designed principally for use in non-domestic premises, for example, machine tools, laundry-type washing machines.

125. **Manufacturing industry**: All activities included within section C. Both cottage industry and large-scale activities are included. It should be noted that the use of heavy plant or machinery is not exclusive to section C.

126. **Product:** A product is the outcome of economic activity. It is the generic term applied to goods and services.

127. **Finished product:** Products for which processing has been completed.

128. **Semi-finished product**: Products that have undergone some processing but require further processing before they are ready for use. They may be sold to other manufacturers or transferred to subcontractors for further processing. Typical examples would include rough metal castings sold or transferred for finishing elsewhere.

129. **Production:** Production is an activity resulting in a product. It is used with reference to the whole range of economic activities. The term is not reserved for the agricultural, mining or manufacturing sectors. It is also used in relation to the service sector. More specific terms may be used to denote production: provision of services, processing, manufacturing, etc., depending on the branch of activity. Production may be measured in various ways either in physical terms or by value.

130. **Transformation:** Transformation is a process that modifies the nature, composition or form of raw materials, semi-finished or finished products for the purpose of obtaining new products.

131. **Treatment:** A process that is carried out, inter alia, for the purpose of protecting certain products, for giving them certain properties or for preventing any harmful effects that might otherwise result from their use. Examples are the treatment of crops, wood, metals and waste.

132. **Value added**: The gross value added at basic price is defined as the difference between output at basic prices and intermediate consumption at purchasers' prices.

Division	Group	Class and Subclass	Description
Section A			**Agriculture, Forestry and Fishing**
01			**Crop and animal production, hunting and related service activities**
	01.1		Growing of non-perennial crops
		01.11	Growing of cereals (except rice), leguminous crops and oil seeds
		01.12	Growing of rice
		01.13	Growing of vegetables and melons, roots and tubers
		01.14	Growing of sugar cane
		01.15	Growing of tobacco
		01.16	Growing of fibre crops
		01.19	Growing of other non-perennial crops
	01.2		Growing of perennial crops
		01.21	Growing of grapes
		01.22	Growing of tropical and subtropical fruits
		01.23	Growing of citrus fruits
		01.24	Growing of pome fruits and stone fruits
		01.25	Growing of other tree and bush fruits and nuts
		01.26	Growing of oleaginous fruits
		01.27	Growing of beverage crops
		01.28	Growing of spices, aromatic, drug and pharmaceutical crops
		01.29	Growing of other perennial crops
	01.3		Plant propagation
		01.30	Plant propagation
	01.4		Animal production
		01.41	Raising of dairy cattle
		01.42	Raising of other cattle and buffaloes
		01.43	Raising of horses and other equines
		01.44	Raising of camels and camelids
		01.45	Raising of sheep and goats
		01.46	Raising of swine/pigs
		01.47	Raising of poultry
		01.49	Raising of other animals
	01.5		Mixed farming
		01.50	Mixed farming
	01.6		Support activities to agriculture and post-harvest crop activities
		01.61	Support activities for crop production
		01.62	Support activities for animal production
		01.62/1	Farm animal boarding and care
		01.62/9	Support activities for animal production (other than farm animal boarding and care) n.e.c.
		01.63	Post-harvest crop activities
		01.64	Seed processing for propagation
	01.7		Hunting, trapping and related service activities
		01.70	Hunting, trapping and related service activities

A
B

Division	Group	Class and Subclass	Description
02			**Forestry and logging**
	02.1		Silviculture and other forestry activities
		02.10	Silviculture and other forestry activities
	02.2		Logging
		02.20	Logging
	02.3		Gathering of wild growing non-wood products
		02.30	Gathering of wild growing non-wood products
	02.4		Support services to forestry
		02.40	Support services to forestry
03			**Fishing and aquaculture**
	03.1		Fishing
		03.11	Marine fishing
		03.12	Freshwater fishing
	03.2		Aquaculture
		03.21	Marine aquaculture
		03.22	Freshwater aquaculture
Section B			**Mining and Quarrying**
05			**Mining of coal and lignite**
	05.1		Mining of hard coal
		05.10	Mining of hard coal
		05.10/1	Mining of hard coal from deep coal mines (underground mining)
		05.10/2	Mining of hard coal from open cast coal working (surface mining)
	05.2		Mining of lignite
		05.20	Mining of lignite
06			**Extraction of crude petroleum and natural gas**
	06.1		Extraction of crude petroleum
		06.10	Extraction of crude petroleum
	06.2		Extraction of natural gas
		06.20	Extraction of natural gas
07			**Mining of metal ores**
	07.1		Mining of iron ores
		07.10	Mining of iron ores
	07.2		Mining of non-ferrous metal ores
		07.21	Mining of uranium and thorium ores
		07.29	Mining of other non-ferrous metal ores
08			**Other mining and quarrying**
	08.1		Quarrying of stone, sand and clay
		08.11	Quarrying of ornamental and building stone, limestone, gypsum, chalk and slate
		08.12	Operation of gravel and sand pits; mining of clays and kaolin
	08.9		Mining and quarrying n.e.c.
		08.91	Mining of chemical and fertiliser minerals

Division	Group	Class and Subclass	Description
		08.92	Extraction of peat
		08.93	Extraction of salt
		08.99	Other mining and quarrying n.e.c.
09			**Mining support service activities**
	09.1		Support activities for petroleum and natural gas extraction
		09.10	Support activities for petroleum and natural gas extraction
	09.9		Support activities for other mining and quarrying
		09.90	Support activities for other mining and quarrying
Section C			**Manufacturing**
10			**Manufacture of food products**
	10.1		Processing and preserving of meat and production of meat products
		10.11	Processing and preserving of meat
		10.12	Processing and preserving of poultry meat
		10.13	Production of meat and poultry meat products
	10.2		Processing and preserving of fish, crustaceans and molluscs
		10.20	Processing and preserving of fish, crustaceans and molluscs
	10.3		Processing and preserving of fruit and vegetables
		10.31	Processing and preserving of potatoes
		10.32	Manufacture of fruit and vegetable juice
		10.39	Other processing and preserving of fruit and vegetables
	10.4		Manufacture of vegetable and animal oils and fats
		10.41	Manufacture of oils and fats
		10.42	Manufacture of margarine and similar edible fats
	10.5		Manufacture of dairy products
		10.51	Operation of dairies and cheese making
		10.51/1	Liquid milk and cream production
		10.51/2	Butter and cheese production
		10.51/9	Manufacture of milk products (other than liquid milk and cream, butter, cheese) n.e.c.
		10.52	Manufacture of ice cream
	10.6		Manufacture of grain mill products, starches and starch products
		10.61	Manufacture of grain mill products
		10.61/1	Grain milling
		10.61/2	Manufacture of breakfast cereals and cereals-based foods
		10.62	Manufacture of starches and starch products
	10.7		Manufacture of bakery and farinaceous products
		10.71	Manufacture of bread; manufacture of fresh pastry goods and cakes
		10.72	Manufacture of rusks and biscuits; manufacture of preserved pastry goods and cakes
		10.73	Manufacture of macaroni, noodles, couscous and similar farinaceous products
	10.8		Manufacture of other food products
		10.81	Manufacture of sugar
		10.82	Manufacture of cocoa, chocolate and sugar confectionery
		10.82/1	Manufacture of cocoa, and chocolate confectionery
		10.82/2	Manufacture of sugar confectionery
		10.83	Processing of tea and coffee
		10.83/1	Tea processing

B

C

Division	Group	Class and Subclass	Description
		10.83/2	Production of coffee and coffee substitutes
		10.84	Manufacture of condiments and seasonings
		10.85	Manufacture of prepared meals and dishes
		10.86	Manufacture of homogenised food preparations and dietetic food
		10.89	Manufacture of other food products n.e.c.
	10.9		Manufacture of prepared animal feeds
		10.91	Manufacture of prepared feeds for farm animals
		10.92	Manufacture of prepared pet foods
11			**Manufacture of beverages**
	11.0		Manufacture of beverages
		11.01	Distilling, rectifying and blending of spirits
		11.02	Manufacture of wine from grape
		11.03	Manufacture of cider and other fruit wines
		11.04	Manufacture of other non-distilled fermented beverages
		11.05	Manufacture of beer
		11.06	Manufacture of malt
		11.07	Manufacture of soft drinks; production of mineral waters and other bottled waters
12			**Manufacture of tobacco products**
	12.0		Manufacture of tobacco products
		12.00	Manufacture of tobacco products
13			**Manufacture of textiles**
	13.1		Preparation and spinning of textile fibres
		13.10	Preparation and spinning of textile fibres
	13.2		Weaving of textiles
		13.20	Weaving of textiles
	13.3		Finishing of textiles
		13.30	Finishing of textiles
	13.9		Manufacture of other textiles
		13.91	Manufacture of knitted and crocheted fabrics
		13.92	Manufacture of made-up textile articles, except apparel
		13.92/1	Manufacture of soft furnishings
		13.92/2	Manufacture of canvas goods, sacks etc.
		13.92/3	Manufacture of household textiles (other than soft furnishings of 13.92/1)
		13.93	Manufacture of carpets and rugs
		13.93/1	Manufacture of woven or tufted carpets and rugs
		13.93/9	Manufacture of carpets and rugs (other than woven or tufted) n.e.c.
		13.94	Manufacture of cordage, rope, twine and netting
		13.95	Manufacture of non-wovens and articles made from non-wovens, except apparel
		13.96	Manufacture of other technical and industrial textiles
		13.99	Manufacture of other textiles n.e.c.
14			**Manufacture of wearing apparel**
	14.1		Manufacture of wearing apparel, except fur apparel
		14.11	Manufacture of leather clothes
		14.12	Manufacture of workwear
		14.13	Manufacture of other outerwear
		14.13/1	Manufacture of men's outerwear, other than leather clothes and workwear
		14.13/2	Manufacture of women's outerwear, other than leather clothes and workwear

Division	Group	Class and Subclass	Description
		14.14	Manufacture of underwear
		14.14/1	Manufacture of men's underwear
		14.14/2	Manufacture of women's underwear
		14.19	Manufacture of other wearing apparel and accessories
	14.2		Manufacture of articles of fur
		14.20	Manufacture of articles of fur
	14.3		Manufacture of knitted and crocheted apparel
		14.31	Manufacture of knitted and crocheted hosiery
		14.39	Manufacture of other knitted and crocheted apparel
15			**Manufacture of leather and related products**
	15.1		Tanning and dressing of leather; manufacture of luggage, handbags, saddlery and harness; dressing and dyeing of fur
		15.11	Tanning and dressing of leather; dressing and dyeing of fur
		15.12	Manufacture of luggage, handbags and the like, saddlery and harness
	15.2		Manufacture of footwear
		15.20	Manufacture of footwear
16			**Manufacture of wood and of products of wood and cork, except furniture; manufacture of articles of straw and plaiting materials**
	16.1		Sawmilling and planing of wood
		16.10	Sawmilling and planing of wood
	16.2		Manufacture of products of wood, cork, straw and plaiting materials
		16.21	Manufacture of veneer sheets and wood-based panels
		16.22	Manufacture of assembled parquet floors
		16.23	Manufacture of other builders' carpentry and joinery
		16.24	Manufacture of wooden containers
		16.29	Manufacture of other products of wood; manufacture of articles of cork, straw and plaiting materials
17			**Manufacture of paper and paper products**
	17.1		Manufacture of pulp, paper and paperboard
		17.11	Manufacture of pulp
		17.12	Manufacture of paper and paperboard
	17.2		Manufacture of articles of paper and paperboard
		17.21	Manufacture of corrugated paper and paperboard and of containers of paper and paperboard
		17.21/1	Manufacture of corrugated paper and paperboard; manufacture of sacks and bags of paper
		17.21/9	Manufacture of paper and paperboard containers other than sacks and bags
		17.22	Manufacture of household and sanitary goods and of toilet requisites
		17.23	Manufacture of paper stationery
		17.24	Manufacture of wallpaper
		17.29	Manufacture of other articles of paper and paperboard
18			**Printing and reproduction of recorded media**
	18.1		Printing and service activities related to printing
		18.11	Printing of newspapers
		18.12	Other printing
		18.12/1	Manufacture of printed labels
		18.12/9	Printing (other than printing of newspapers and printing on labels and tags) n.e.c.
		18.13	Pre-press and pre-media services
		18.14	Binding and related services

C

31

Division	Group	Class and Subclass	Description
	18.2		Reproduction of recorded media
		18.20	Reproduction of recorded media
		18.20/1	Reproduction of sound recording
		18.20/2	Reproduction of video recording
		18.20/3	Reproduction of computer media
19			**Manufacture of coke and refined petroleum products**
	19.1		Manufacture of coke oven products
		19.10	Manufacture of coke oven products
	19.2		Manufacture of refined petroleum products
		19.20	Manufacture of refined petroleum products
		19.20/1	Mineral oil refining
		19.20/9	Other treatment of petroleum products (excluding mineral oil refining/petrochemicals manufacture)
20			**Manufacture of chemicals and chemical products**
	20.1		Manufacture of basic chemicals, fertilisers and nitrogen compounds, plastics and synthetic rubber in primary forms
		20.11	Manufacture of industrial gases
		20.12	Manufacture of dyes and pigments
		20.13	Manufacture of other inorganic basic chemicals
		20.14	Manufacture of other organic basic chemicals
		20.15	Manufacture of fertilisers and nitrogen compounds
		20.16	Manufacture of plastics in primary forms
		20.17	Manufacture of synthetic rubber in primary forms
	20.2		Manufacture of pesticides and other agrochemical products
		20.20	Manufacture of pesticides and other agrochemical products
	20.3		Manufacture of paints, varnishes and similar coatings, printing ink and mastics
		20.30	Manufacture of paints, varnishes and similar coatings, printing ink and mastics
		20.30/1	Manufacture of paints, varnishes and similar coatings, mastics and sealants
		20.30/2	Manufacture of printing ink
	20.4		Manufacture of soap and detergents, cleaning and polishing preparations, perfumes and toilet preparations
		20.41	Manufacture of soap and detergents, cleaning and polishing preparations
		20.41/1	Manufacture of soap and detergents
		20.41/2	Manufacture of cleaning and polishing preparations
		20.42	Manufacture of perfumes and toilet preparations
	20.5		Manufacture of other chemical products
		20.51	Manufacture of explosives
		20.52	Manufacture of glues
		20.53	Manufacture of essential oils
		20.59	Manufacture of other chemical products n.e.c.
	20.6		Manufacture of man-made fibres
		20.60	Manufacture of man-made fibres
21			**Manufacture of basic pharmaceutical products and pharmaceutical preparations**
	21.1		Manufacture of basic pharmaceutical products
		21.10	Manufacture of basic pharmaceutical products
	21.2		Manufacture of pharmaceutical preparations
		21.20	Manufacture of pharmaceutical preparations

Division	Group	Class and Subclass	Description
22			**Manufacture of rubber and plastic products**
	22.1		Manufacture of rubber products
		22.11	Manufacture of rubber tyres and tubes; retreading and rebuilding of rubber tyres
		22.19	Manufacture of other rubber products
	22.2		Manufacture of plastics products
		22.21	Manufacture of plastic plates, sheets, tubes and profiles
		22.22	Manufacture of plastic packing goods
		22.23	Manufacture of builders' ware of plastic
		22.29	Manufacture of other plastic products
23			**Manufacture of other non-metallic mineral products**
	23.1		Manufacture of glass and glass products
		23.11	Manufacture of flat glass
		23.12	Shaping and processing of flat glass
		23.13	Manufacture of hollow glass
		23.14	Manufacture of glass fibres
		23.19	Manufacture and processing of other glass, including technical glassware
	23.2		Manufacture of refractory products
		23.20	Manufacture of refractory products
	23.3		Manufacture of clay building materials
		23.31	Manufacture of ceramic tiles and flags
		23.32	Manufacture of bricks, tiles and construction products, in baked clay
	23.4		Manufacture of other porcelain and ceramic products
		23.41	Manufacture of ceramic household and ornamental articles
		23.42	Manufacture of ceramic sanitary fixtures
		23.43	Manufacture of ceramic insulators and insulating fittings
		23.44	Manufacture of other technical ceramic products
		23.49	Manufacture of other ceramic products
	23.5		Manufacture of cement, lime and plaster
		23.51	Manufacture of cement
		23.52	Manufacture of lime and plaster
	23.6		Manufacture of articles of concrete, cement and plaster
		23.61	Manufacture of concrete products for construction purposes
		23.62	Manufacture of plaster products for construction purposes
		23.63	Manufacture of ready-mixed concrete
		23.64	Manufacture of mortars
		23.65	Manufacture of fibre cement
		23.69	Manufacture of other articles of concrete, plaster and cement
	23.7		Cutting, shaping and finishing of stone
		23.70	Cutting, shaping and finishing of stone
	23.9		Manufacture of abrasive products and non-metallic mineral products n.e.c.
		23.91	Production of abrasive products
		23.99	Manufacture of other non-metallic mineral products n.e.c.
24			**Manufacture of basic metals**
	24.1		Manufacture of basic iron and steel and of ferro-alloys
		24.10	Manufacture of basic iron and steel and of ferro-alloys

Division	Group	Class and Subclass	Description
	24.2		Manufacture of tubes, pipes, hollow profiles and related fittings, of steel
		24.20	Manufacture of tubes, pipes, hollow profiles and related fittings, of steel
	24.3		Manufacture of other products of first processing of steel
		24.31	Cold drawing of bars
		24.32	Cold rolling of narrow strip
		24.33	Cold forming or folding
		24.34	Cold drawing of wire
	24.4		Manufacture of basic precious and other non-ferrous metals
		24.41	Precious metals production
		24.42	Aluminium production
		24.43	Lead, zinc and tin production
		24.44	Copper production
		24.45	Other non-ferrous metal production
		24.46	Processing of nuclear fuel
	24.5		Casting of metals
		24.51	Casting of iron
		24.52	Casting of steel
		24.53	Casting of light metals
		24.54	Casting of other non-ferrous metals
25			**Manufacture of fabricated metal products, except machinery and equipment**
	25.1		Manufacture of structural metal products
		25.11	Manufacture of metal structures and parts of structures
		25.12	Manufacture of doors and windows of metal
	25.2		Manufacture of tanks, reservoirs and containers of metal
		25.21	Manufacture of central heating radiators and boilers
		25.29	Manufacture of other tanks, reservoirs and containers of metal
	25.3		Manufacture of steam generators, except central heating hot water boilers
		25.30	Manufacture of steam generators, except central heating hot water boilers
	25.4		Manufacture of weapons and ammunition
		25.40	Manufacture of weapons and ammunition
	25.5		Forging, pressing, stamping and roll-forming of metal; powder metallurgy
		25.50	Forging, pressing, stamping and roll-forming of metal; powder metallurgy
	25.6		Treatment and coating of metals; machining
		25.61	Treatment and coating of metals
		25.62	Machining
	25.7		Manufacture of cutlery, tools and general hardware
		25.71	Manufacture of cutlery
		25.72	Manufacture of locks and hinges
		25.73	Manufacture of tools
	25.9		Manufacture of other fabricated metal products
		25.91	Manufacture of steel drums and similar containers
		25.92	Manufacture of light metal packaging
		25.93	Manufacture of wire products, chain and springs
		25.94	Manufacture of fasteners and screw machine products
		25.99	Manufacture of other fabricated metal products n.e.c.

Division	Group	Class and Subclass	Description
26			**Manufacture of computer, electronic and optical products**
	26.1		Manufacture of electronic components and boards
		26.11	Manufacture of electronic components
		26.12	Manufacture of loaded electronic boards
	26.2		Manufacture of computers and peripheral equipment
		26.20	Manufacture of computers and peripheral equipment
	26.3		Manufacture of communication equipment
		26.30	Manufacture of communication equipment
		26.30/1	Manufacture of telegraph and telephone apparatus and equipment
		26.30/9	Manufacture of communication equipment (other than telegraph and telephone apparatus and equipment)
	26.4		Manufacture of consumer electronics
		26.40	Manufacture of consumer electronics
	26.5		Manufacture of instruments and appliances for measuring, testing and navigation; watches and clocks
		26.51	Manufacture of instruments and appliances for measuring, testing and navigation
		26.51/1	Manufacture of electronic instruments and appliances for measuring, testing, and navigation, except industrial process control equipment
		26.51/2	Manufacture of electronic industrial process control equipment
		26.51/3	Manufacture of non-electronic instruments and appliances for measuring, testing and navigation, except industrial process control equipment
		26.51/4	Manufacture of non-electronic industrial process control equipment
		26.52	Manufacture of watches and clocks
	26.6		Manufacture of irradiation, electromedical and electrotherapeutic equipment
		26.60	Manufacture of irradiation, electromedical and electrotherapeutic equipment
	26.7		Manufacture of optical instruments and photographic equipment
		26.70	Manufacture of optical instruments and photographic equipment
		26.70/1	Manufacture of optical precision instruments
		26.70/2	Manufacture of photographic and cinematographic equipment
	26.8		Manufacture of magnetic and optical media
		26.80	Manufacture of magnetic and optical media
27			**Manufacture of electrical equipment**
	27.1		Manufacture of electric motors, generators, transformers and electricity distribution and control apparatus
		27.11	Manufacture of electric motors, generators and transformers
		27.12	Manufacture of electricity distribution and control apparatus
	27.2		Manufacture of batteries and accumulators
		27.20	Manufacture of batteries and accumulators
	27.3		Manufacture of wiring and wiring devices
		27.31	Manufacture of fibre optic cables
		27.32	Manufacture of other electronic and electric wires and cables
		27.33	Manufacture of wiring devices
	27.4		Manufacture of electric lighting equipment
		27.40	Manufacture of electric lighting equipment
	27.5		Manufacture of domestic appliances
		27.51	Manufacture of electric domestic appliances
		27.52	Manufacture of non-electric domestic appliances

C

Division	Group	Class and Subclass	Description
	27.9		Manufacture of other electrical equipment
		27.90	Manufacture of other electrical equipment
28			**Manufacture of machinery and equipment n.e.c.**
	28.1		Manufacture of general-purpose machinery
		28.11	Manufacture of engines and turbines, except aircraft, vehicle and cycle engines
		28.12	Manufacture of fluid power equipment
		28.13	Manufacture of other pumps and compressors
		28.13/1	Manufacture of pumps
		28.13/2	Manufacture of compressors
		28.14	Manufacture of other taps and valves
		28.15	Manufacture of bearings, gears, gearing and driving elements
	28.2		Manufacture of other general-purpose machinery
		28.21	Manufacture of ovens, furnaces and furnace burners
		28.22	Manufacture of lifting and handling equipment
		28.23	Manufacture of office machinery and equipment (except computers and peripheral equipment)
		28.24	Manufacture of power-driven hand tools
		28.25	Manufacture of non-domestic cooling and ventilation equipment
		28.29	Manufacture of other general-purpose machinery n.e.c.
	28.3		Manufacture of agricultural and forestry machinery
		28.30	Manufacture of agricultural and forestry machinery
		28.30/1	Manufacture of agricultural tractors
		28.30/2	Manufacture of agricultural and forestry machinery (other than agricultural tractors)
	28.4		Manufacture of metal forming machinery and machine tools
		28.41	Manufacture of metal forming machinery
		28.49	Manufacture of other machine tools
	28.9		Manufacture of other special-purpose machinery
		28.91	Manufacture of machinery for metallurgy
		28.92	Manufacture of machinery for mining, quarrying and construction
		28.92/1	Manufacture of machinery for mining
		28.92/2	Manufacture of earthmoving equipment
		28.92/3	Manufacture of equipment for concrete crushing and screening roadworks
		28.93	Manufacture of machinery for food, beverage and tobacco processing
		28.94	Manufacture of machinery for textile, apparel and leather production
		28.95	Manufacture of machinery for paper and paperboard production
		28.96	Manufacture of plastics and rubber machinery
		28.99	Manufacture of other special-purpose machinery n.e.c.
29			**Manufacture of motor vehicles, trailers and semi-trailers**
	29.1		Manufacture of motor vehicles
		29.10	Manufacture of motor vehicles
	29.2		Manufacture of bodies (coachwork) for motor vehicles; manufacture of trailers and semi-trailers
		29.20	Manufacture of bodies (coachwork) for motor vehicles; manufacture of trailers and semi-trailers
		29.20/1	Manufacture of bodies (coachwork) for motor vehicles (except caravans)
		29.20/2	Manufacture of trailers and semi-trailers
		29.20/3	Manufacture of caravans
	29.3		Manufacture of parts and accessories for motor vehicles
		29.31	Manufacture of electrical and electronic equipment for motor vehicles
		29.32	Manufacture of other parts and accessories for motor vehicles

C

Division	Group	Class and Subclass	Description
30			**Manufacture of other transport equipment**
	30.1		Building of ships and boats
		30.11	Building of ships and floating structures
		30.12	Building of pleasure and sporting boats
	30.2		Manufacture of railway locomotives and rolling stock
		30.20	Manufacture of railway locomotives and rolling stock
	30.3		Manufacture of air and spacecraft and related machinery
		30.30	Manufacture of air and spacecraft and related machinery
	30.4		Manufacture of military fighting vehicles
		30.40	Manufacture of military fighting vehicles
	30.9		Manufacture of transport equipment n.e.c.
		30.91	Manufacture of motorcycles
		30.92	Manufacture of bicycles and invalid carriages
		30.99	Manufacture of other transport equipment n.e.c.
31			**Manufacture of furniture**
	31.0		Manufacture of furniture
		31.01	Manufacture of office and shop furniture
		31.02	Manufacture of kitchen furniture
		31.03	Manufacture of mattresses
		31.09	Manufacture of other furniture
32			**Other manufacturing**
	32.1		Manufacture of jewellery, bijouterie and related articles
		32.11	Striking of coins
		32.12	Manufacture of jewellery and related articles
		32.13	Manufacture of imitation jewellery and related articles
	32.2		Manufacture of musical instruments
		32.20	Manufacture of musical instruments
	32.3		Manufacture of sports goods
		32.30	Manufacture of sports goods
	32.4		Manufacture of games and toys
		32.40	Manufacture of games and toys
		32.40/1	Manufacture of professional and arcade games and toys
		32.40/9	Manufacture of games and toys (other than professional and arcade games and toys) n.e.c.
	32.5		Manufacture of medical and dental instruments and supplies
		32.50	Manufacture of medical and dental instruments and supplies
	32.9		Manufacturing n.e.c.
		32.91	Manufacture of brooms and brushes
		32.99	Other manufacturing n.e.c.
33			**Repair and installation of machinery and equipment**
	33.1		Repair of fabricated metal products, machinery and equipment
		33.11	Repair of fabricated metal products
		33.12	Repair of machinery

Division	Group	Class and Subclass	Description
		33.13	Repair of electronic and optical equipment
		33.14	Repair of electrical equipment
		33.15	Repair and maintenance of ships and boats
		33.16	Repair and maintenance of aircraft and spacecraft
		33.17	Repair and maintenance of other transport equipment
		33.19	Repair of other equipment
	33.2		Installation of industrial machinery and equipment
		33.20	Installation of industrial machinery and equipment

Section D   **Electricity, Gas, Steam and Air Conditioning Supply**

35   **Electricity, gas, steam and air conditioning supply**

Division	Group	Class and Subclass	Description
	35.1		Electric power generation, transmission and distribution
		35.11	Production of electricity
		35.12	Transmission of electricity
		35.13	Distribution of electricity
		35.14	Trade of electricity
	35.2		Manufacture of gas; distribution of gaseous fuels through mains
		35.21	Manufacture of gas
		35.22	Distribution of gaseous fuels through mains
		35.23	Trade of gas through mains
	35.3		Steam and air conditioning supply
		35.30	Steam and air conditioning supply

Section E   **Water Supply; Sewerage, Waste Management and Remediation Activities**

36   **Water collection, treatment and supply**

Division	Group	Class and Subclass	Description
	36.0		Water collection, treatment and supply
		36.00	Water collection, treatment and supply

37   **Sewerage**

Division	Group	Class and Subclass	Description
	37.0		Sewerage
		37.00	Sewerage

38   **Waste collection, treatment and disposal activities; materials recovery**

Division	Group	Class and Subclass	Description
	38.1		Waste collection
		38.11	Collection of non-hazardous waste
		38.12	Collection of hazardous waste
	38.2		Waste treatment and disposal
		38.21	Treatment and disposal of non-hazardous waste
		38.22	Treatment and disposal of hazardous waste
	38.3		Materials recovery
		38.31	Dismantling of wrecks
		38.32	Recovery of sorted materials

39   **Remediation activities and other waste management services**

This division includes the provision of remediation services, i.e. the cleanup of contaminated buildings and sites, soil, surface or ground water.

Division	Group	Class and Subclass	Description
	39.0		Remediation activities and other waste management services
		39.00	Remediation activities and other waste management services

Division	Group	Class and Subclass	Description
Section F			**Construction**
41			**Construction of buildings**
	41.1		Development of building projects
		41.10	Development of building projects
	41.2		Construction of residential and non-residential buildings
		41.20	Construction of residential and non-residential buildings
		41.20/1	Construction of commercial buildings
		41.20/2	Construction of domestic buildings
42			**Civil engineering**
	42.1		Construction of roads and railways
		42.11	Construction of roads and motorways
		42.12	Construction of railways and underground railways
		42.13	Construction of bridges and tunnels
	42.2		Construction of utility projects
		42.21	Construction of utility projects for fluids
		42.22	Construction of utility projects for electricity and telecommunications
	42.9		Construction of other civil engineering projects
		42.91	Construction of water projects
		42.99	Construction of other civil engineering projects n.e.c.
43			**Specialised construction activities**
	43.1		Demolition and site preparation
		43.11	Demolition
		43.12	Site preparation
		43.13	Test drilling and boring
	43.2		Electrical, plumbing and other construction installation activities
		43.21	Electrical installation
		43.22	Plumbing, heat and air-conditioning installation
		43.29	Other construction installation
	43.3		Building completion and finishing
		43.31	Plastering
		43.32	Joinery installation
		43.33	Floor and wall covering
		43.34	Painting and glazing
		43.34/1	Painting
		43.34/2	Glazing
		43.39	Other building completion and finishing
	43.9		Other specialised construction activities
		43.91	Roofing activities
		43.99	Other specialised construction activities n.e.c.
		43.99/1	Scaffold erection
		43.99/9	Specialised construction activities (other than scaffold erection) n.e.c.
Section G			**Wholesale and Retail Trade; Repair of Motor Vehicles and Motorcycles**
45			**Wholesale and retail trade and repair of motor vehicles and motorcycles**
	45.1		Sale of motor vehicles
		45.11	Sale of cars and light motor vehicles

F
G

Division	Group	Class and Subclass	Description
		45.11/1	Sale of new cars and light motor vehicles
		45.11/2	Sale of used cars and light motor vehicles
		45.19	Sale of other motor vehicles
	45.2		Maintenance and repair of motor vehicles
		45.20	Maintenance and repair of motor vehicles
	45.3		Sale of motor vehicle parts and accessories
		45.31	Wholesale trade of motor vehicle parts and accessories
		45.32	Retail trade of motor vehicle parts and accessories
	45.4		Sale, maintenance and repair of motorcycles and related parts and accessories
		45.40	Sale, maintenance and repair of motorcycles and related parts and accessories
46			**Wholesale trade, except of motor vehicles and motorcycles**
	46.1		Wholesale on a fee or contract basis
		46.11	Agents involved in the sale of agricultural raw materials, live animals, textile raw materials and semi-finished goods
		46.12	Agents involved in the sale of fuels, ores, metals and industrial chemicals
		46.13	Agents involved in the sale of timber and building materials
		46.14	Agents involved in the sale of machinery, industrial equipment, ships and aircraft
		46.15	Agents involved in the sale of furniture, household goods, hardware and ironmongery
		46.16	Agents involved in the sale of textiles, clothing, fur, footwear and leather goods
		46.17	Agents involved in the sale of food, beverages and tobacco
		46.18	Agents specialised in the sale of other particular products
		46.19	Agents involved in the sale of a variety of goods
	46.2		Wholesale of agricultural raw materials and live animals
		46.21	Wholesale of grain, unmanufactured tobacco, seeds and animal feeds
		46.22	Wholesale of flowers and plants
		46.23	Wholesale of live animals
		46.24	Wholesale of hides, skins and leather
	46.3		Wholesale of food, beverages and tobacco
		46.31	Wholesale of fruit and vegetables
		46.32	Wholesale of meat and meat products
		46.33	Wholesale of dairy products, eggs and edible oils and fats
		46.34	Wholesale of beverages
		46.34/1	Wholesale of fruit and vegetable juices, mineral waters and soft drinks
		46.34/2	Wholesale of wine, beer, spirits and other alcoholic beverages
		46.35	Wholesale of tobacco products
		46.36	Wholesale of sugar and chocolate and sugar confectionery
		46.37	Wholesale of coffee, tea, cocoa and spices
		46.38	Wholesale of other food, including fish, crustaceans and molluscs
		46.39	Non-specialised wholesale of food, beverages and tobacco
	46.4		Wholesale of household goods
		46.41	Wholesale of textiles
		46.42	Wholesale of clothing and footwear
		46.43	Wholesale of electrical household appliances
		46.43/1	Wholesale of gramophone records, audio tapes, compact discs and video tapes and of the equipment on which these are played
		46.43/9	Wholesale of radio and television goods and of electrical household appliances (other than of gramophone records, audio tapes, compact discs and video tapes and the equipment on which these are played) n.e.c.
		46.44	Wholesale of china and glassware and cleaning materials
		46.45	Wholesale of perfume and cosmetics

G

Division	Group	Class and Subclass	Description
		46.46	Wholesale of pharmaceutical goods
		46.47	Wholesale of furniture, carpets and lighting equipment
		46.48	Wholesale of watches and jewellery
		46.49	Wholesale of other household goods
		46.49/1	Wholesale of musical instruments
		46.49/9	Wholesale of household goods (other than musical instruments) n.e.c.
	46.5		Wholesale of information and communication equipment
		46.51	Wholesale of computers, computer peripheral equipment and software
		46.52	Wholesale of electronic and telecommunications equipment and parts
	46.6		Wholesale of other machinery, equipment and supplies
		46.61	Wholesale of agricultural machinery, equipment and supplies
		46.62	Wholesale of machine tools
		46.63	Wholesale of mining, construction and civil engineering machinery
		46.64	Wholesale of machinery for the textile industry and of sewing and knitting machines
		46.65	Wholesale of office furniture
		46.66	Wholesale of other office machinery and equipment
		46.69	Wholesale of other machinery and equipment
	46.7		Other specialised wholesale
		46.71	Wholesale of solid, liquid and gaseous fuels and related products
		46.71/1	Wholesale of petroleum and petroleum products
		46.71/9	Wholesale of fuels and related products (other than petroleum and petroleum products)
		46.72	Wholesale of metals and metal ores
		46.73	Wholesale of wood, construction materials and sanitary equipment
		46.74	Wholesale of hardware, plumbing and heating equipment and supplies
		46.75	Wholesale of chemical products
		46.76	Wholesale of other intermediate products
		46.77	Wholesale of waste and scrap
	46.9		Non-specialised wholesale trade
		46.90	Non-specialised wholesale trade
47			**Retail trade, except of motor vehicles and motorcycles**
	47.1		Retail sale in non-specialised stores
		47.11	Retail sale in non-specialised stores with food, beverages or tobacco predominating
		47.19	Other retail sale in non-specialised stores
	47.2		Retail sale of food, beverages and tobacco in specialised stores
		47.21	Retail sale of fruit and vegetables in specialised stores
		47.22	Retail sale of meat and meat products in specialised stores
		47.23	Retail sale of fish, crustaceans and molluscs in specialised stores
		47.24	Retail sale of bread, cakes, flour confectionery and sugar confectionery in specialised stores
		47.25	Retail sale of beverages in specialised stores
		47.26	Retail sale of tobacco products in specialised stores
		47.29	Other retail sale of food in specialised stores
	47.3		Retail sale of automotive fuel in specialised stores
		47.30	Retail sale of automotive fuel in specialised stores
	47.4		Retail sale of information and communication equipment in specialised stores
		47.41	Retail sale of computers, peripheral units and software in specialised stores
		47.42	Retail sale of telecommunications equipment in specialised stores
		47.42/1	Retail sale of mobile telephones in specialised stores

G

Division	Group	Class and Subclass	Description
		47.42/9	Retail sale of telecommunications equipment (other than mobile telephones) n.e.c., in specialised stores
		47.43	Retail sale of audio and video equipment in specialised stores
	47.5		Retail sale of other household equipment in specialised stores
		47.51	Retail sale of textiles in specialised stores
		47.52	Retail sale of hardware, paints and glass in specialised stores
		47.53	Retail sale of carpets, rugs, wall and floor coverings in specialised stores
		47.54	Retail sale of electrical household appliances in specialised stores
		47.59	Retail sale of furniture, lighting equipment and other household articles in specialised stores
		47.59/1	Retail sale of musical instruments and scores in specialised stores
		47.59/9	Retail sale of furniture, lighting equipment and other household articles (other than musical instruments) n.e.c., in specialised stores
	47.6		Retail sale of cultural and recreation goods in specialised stores
		47.61	Retail sale of books in specialised stores
		47.62	Retail sale of newspapers and stationery in specialised stores
		47.63	Retail sale of music and video recordings in specialised stores
		47.64	Retail sale of sporting equipment in specialised stores
		47.65	Retail sale of games and toys in specialised stores
	47.7		Retail sale of other goods in specialised stores
		47.71	Retail sale of clothing in specialised stores
		47.72	Retail sale of footwear and leather goods in specialised stores
		47.72/1	Retail sale of footwear in specialised stores
		47.72/2	Retail sale of leather goods in specialised stores
		47.73	Dispensing chemist in specialised stores
		47.74	Retail sale of medical and orthopaedic goods in specialised stores
		47.74/1	Retail sale of hearing aids in specialised stores
		47.74/9	Retail sale of medical and orthopaedic goods (other than hearing aids) n.e.c., in specialised stores
		47.75	Retail sale of cosmetic and toilet articles in specialised stores
		47.76	Retail sale of flowers, plants, seeds, fertilisers, pet animals and pet food in specialised stores
		47.77	Retail sale of watches and jewellery in specialised stores
		47.78	Other retail sale of new goods in specialised stores
		47.78/1	Retail sale in commercial art galleries
		47.78/2	Retail sale by opticians
		47.78/9	Other retail sale of new goods in specialised stores (other than by opticians or commercial art galleries), n.e.c.
		47.79	Retail sale of second-hand goods in stores
		47.79/1	Retail sale of antiques including antique books, in stores
		47.79/9	Retail sale of second-hand goods (other than antiques and antique books) in stores
	47.8		Retail sale via stalls and markets
		47.81	Retail sale via stalls and markets of food, beverages and tobacco products
		47.82	Retail sale via stalls and markets of textiles, clothing and footwear
		47.89	Retail sale via stalls and markets of other goods
	47.9		Retail trade not in stores, stalls or markets
		47.91	Retail sale via mail order houses or via Internet
		47.99	Other retail sale not in stores, stalls or markets
Section H			**Transportation and Storage**
49			**Land transport and transport via pipelines**
	49.1		Passenger rail transport, interurban
		49.10	Passenger rail transport, interurban
	49.2		Freight rail transport
		49.20	Freight rail transport

Division	Group	Class and Subclass	Description
	49.3		Other passenger land transport
		49.31	Urban and suburban passenger land transport
		49.31/1	Urban, suburban or metropolitan area passenger railway transportation by underground, metro and similar systems
		49.31/9	Urban, suburban or metropolitan area passenger land transport other than railway transportation by underground, metro and similar systems
		49.32	Taxi operation
		49.39	Other passenger land transport n.e.c.
	49.4		Freight transport by road and removal services
		49.41	Freight transport by road
		49.42	Removal services
	49.5		Transport via pipeline
		49.50	Transport via pipeline
50			**Water transport**
	50.1		Sea and coastal passenger water transport
		50.10	Sea and coastal passenger water transport
	50.2		Sea and coastal freight water transport
		50.20	Sea and coastal freight water transport
	50.3		Inland passenger water transport
		50.30	Inland passenger water transport
	50.4		Inland freight water transport
		50.40	Inland freight water transport
51			**Air transport**
	51.1		Passenger air transport
		51.10	Passenger air transport
		51.10/1	Scheduled passenger air transport
		51.10/2	Non-scheduled passenger air transport
	51.2		Freight air transport and space transport
		51.21	Freight air transport
		51.22	Space transport
52			**Warehousing and support activities for transportation**
	52.1		Warehousing and storage
		52.10	Warehousing and storage
		52.10/1	Operation of warehousing and storage facilities for water transport activities of division 50
		52.10/2	Operation of warehousing and storage facilities for air transport activities of division 51
		52.10/3	Operation of warehousing and storage facilities for land transport activities of division 49
	52.2		Support activities for transportation
		52.21	Service activities incidental to land transportation
		52.21/1	Operation of rail freight terminals
		52.21/2	Operation of rail passenger facilities at railway stations
		52.21/3	Operation of bus and coach passenger facilities at bus and coach stations
		52.21/9	Other service activities incidental to land transportation, n.e.c. (not including operation of rail freight terminals, passenger facilities at railway stations or passenger facilities at bus and coach stations)
		52.22	Service activities incidental to water transportation
		52.23	Service activities incidental to air transportation

H

Division	Group	Class and Subclass	Description
		52.24	Cargo handling
		52.24/1	Cargo handling for water transport activities of division 50
		52.24/2	Cargo handling for air transport activities of division 51
		52.24/3	Cargo handling for land transport activities of division 49
		52.29	Other transportation support activities
53			**Postal and courier activities**
	53.1		Postal activities under universal service obligation
		53.10	Postal activities under universal service obligation
	53.2		Other postal and courier activities
		53.20	Other postal and courier activities
		53.20/1	Licensed carriers
		53.20/2	Unlicensed carriers
Section I			**Accommodation and Food Service Activities**
55			**Accommodation**
	55.1		Hotels and similar accommodation
		55.10	Hotels and similar accommodation
	55.2		Holiday and other short-stay accommodation
		55.20	Holiday and other short-stay accommodation
		55.20/1	Holiday centres and villages
		55.20/2	Youth hostels
		55.20/9	Other holiday and other short-stay accommodation (not including holiday centres and villages or youth hostels) n.e.c.
	55.3		Camping grounds, recreational vehicle parks and trailer parks
		55.30	Camping grounds, recreational vehicle parks and trailer parks
	55.9		Other accommodation
		55.90	Other accommodation
56			**Food and beverage service activities**
	56.1		Restaurants and mobile food service activities
		56.10	Restaurants and mobile food service activities
		56.10/1	Licensed restaurants
		56.10/2	Unlicensed restaurants and cafes
		56.10/3	Take away food shops and mobile food stands
	56.2		Event catering and other food service activities
		56.21	Event catering activities
		56.29	Other food service activities
	56.3		Beverage serving activities
		56.30	Beverage serving activities
		56.30/1	Licensed clubs
		56.30/2	Public houses and bars
Section J			**Information and Communication**
58			**Publishing activities**
	58.1		Publishing of books, periodicals and other publishing activities
		58.11	Book publishing
		58.12	Publishing of directories and mailing lists
		58.13	Publishing of newspapers

H
I
J

Division	Group	Class and Subclass	Description
		58.14	Publishing of journals and periodicals
		58.14/1	Publishing of learned journals
		58.14/2	Publishing of consumer, business and professional journals and periodicals
		58.19	Other publishing activities
	58.2		Software publishing
		58.21	Publishing of computer games
		58.29	Other software publishing
59			**Motion picture, video and television programme production, sound recording and music publishing activities**
	59.1		Motion picture, video and television programme activities
		59.11	Motion picture, video and television programme production activities
		59.11/1	Motion picture production activities
		59.11/2	Video production activities
		59.11/3	Television programme production activities
		59.12	Motion picture, video and television programme post-production activities
		59.13	Motion picture, video and television programme distribution activities
		59.13/1	Motion picture distribution activities
		59.13/2	Video distribution activities
		59.13/3	Television programme distribution activities
		59.14	Motion picture projection activities
	59.2		Sound recording and music publishing activities
		59.20	Sound recording and music publishing activities
60			**Programming and broadcasting activities**
	60.1		Radio broadcasting
		60.10	Radio broadcasting
	60.2		Television programming and broadcasting activities
		60.20	Television programming and broadcasting activities
61			**Telecommunications**
	61.1		Wired telecommunications activities
		61.10	Wired telecommunications activities
	61.2		Wireless telecommunications activities
		61.20	Wireless telecommunications activities
	61.3		Satellite telecommunications activities
		61.30	Satellite telecommunications activities
	61.9		Other telecommunications activities
		61.90	Other telecommunications activities
62			**Computer programming, consultancy and related activities**
	62.0		Computer programming, consultancy and related activities
		62.01	Computer programming activities
		62.01/1	Ready-made interactive leisure and entertainment software development
		62.01/2	Business and domestic software development
		62.02	Computer consultancy activities
		62.03	Computer facilities management activities
		62.09	Other information technology and computer service activities

J

Division	Group	Class and Subclass	Description
63			**Information service activities**
	63.1		Data processing, hosting and related activities; web portals
		63.11	Data processing, hosting and related activities
		63.12	Web portals
	63.9		Other information service activities
		63.91	News agency activities
		63.99	Other information service activities n.e.c.
Section K			**Financial and Insurance Activities**
64			**Financial service activities, except insurance and pension funding**
	64.1		Monetary intermediation
		64.11	Central banking
		64.19	Other monetary intermediation
		64.19/1	Banks
		64.19/2	Building societies
	64.2		Activities of holding companies
		64.20	Activities of holding companies
		64.20/1	Activities of agricultural holding companies
		64.20/2	Activities of production holding companies
		64.20/3	Activities of construction holding companies
		64.20/4	Activities of distribution holding companies
		64.20/5	Activities of financial services holding companies
		64.20/9	Activities of other holding companies (not including agricultural, production, construction, distribution and financial services holding companies) n.e.c.
	64.3		Trusts, funds and similar financial entities
		64.30	Trusts, funds and similar financial entities
		64.30/1	Activities of investment trusts
		64.30/2	Activities of unit trusts
		64.30/3	Activities of venture and development capital companies
		64.30/4	Activities of open-ended investment companies
		64.30/5	Activities of property unit trusts
		64.30/6	Activities of real estate investment trusts
	64.9		Other financial service activities, except insurance and pension funding
		64.91	Financial leasing
		64.92	Other credit granting
		64.92/1	Credit granting by non-deposit taking finance houses and other specialist consumer credit grantors
		64.92/2	Activities of mortgage finance companies
		64.92/9	Other credit granting (not including credit granting by non-deposit taking finance houses and other specialist consumer credit grantors and activities of mortgage finance companies) n.e.c.
		64.99	Other financial service activities, except insurance and pension funding, n.e.c.
		64.99/1	Security dealing on own account
		64.99/2	Factoring
		64.99/9	Other financial service activities, except insurance and pension funding, (not including security dealing on own account and factoring) n.e.c.
65			**Insurance, reinsurance and pension funding, except compulsory social security**
	65.1		Insurance
		65.11	Life insurance
		65.12	Non-life insurance
	65.2		Reinsurance
		65.20	Reinsurance

J

K

Division	Group	Class and Subclass	Description
		65.20/1	Life reinsurance
		65.20/2	Non-life reinsurance
	65.3		Pension funding
		65.30	Pension funding
66			**Activities auxiliary to financial services and insurance activities**
	66.1		Activities auxiliary to financial services, except insurance and pension funding
		66.11	Administration of financial markets
		66.12	Security and commodity contracts brokerage
		66.19	Other activities auxiliary to financial services, except insurance and pension funding
	66.2		Activities auxiliary to insurance and pension funding
		66.21	Risk and damage evaluation
		66.22	Activities of insurance agents and brokers
		66.29	Other activities auxiliary to insurance and pension funding
	66.3		Fund management activities
		66.30	Fund management activities
Section L			**Real Estate Activities**
68			**Real estate activities**
	68.1		Buying and selling of own real estate
		68.10	Buying and selling of own real estate
	68.2		Renting and operating of own or leased real estate
		68.20	Renting and operating of own or leased real estate
		68.20/1	Renting and operating of Housing Association real estate
		68.20/2	Letting and operating of conference and exhibition centres
		68.20/9	Letting and operating of own or leased real estate (other than Housing Association real estate and conference and exhibition services) n.e.c.
	68.3		Real estate activities on a fee or contract basis
		68.31	Real estate agencies
		68.32	Management of real estate on a fee or contract basis
Section M			**Professional, Scientific and Technical Activities**
69			**Legal and accounting activities**
	69.1		Legal activities
		69.10	Legal activities
		69.10/1	Barristers at law
		69.10/2	Solicitors
		69.10/9	Activities of patent and copyright agents; other legal activities (other than those of barristers and solicitors) n.e.c.
	69.2		Accounting, bookkeeping and auditing activities; tax consultancy
		69.20	Accounting, bookkeeping and auditing activities; tax consultancy
		69.20/1	Accounting, and auditing activities
		69.20/2	Bookkeeping activities
		69.20/3	Tax consultancy
70			**Activities of head offices; management consultancy activities**
	70.1		Activities of head offices
		70.10	Activities of head offices
	70.2		Management consultancy activities
		70.21	Public relations and communication activities

K
L
M

Division	Group	Class and Subclass	Description
		70.22	Business and other management consultancy activities
		70.22/1	Financial management
		70.22/9	Management consultancy activities (other than financial management)
71			**Architectural and engineering activities; technical testing and analysis**
	71.1		Architectural and engineering activities and related technical consultancy
		71.11	Architectural activities
		71.11/1	Architectural activities
		71.11/2	Urban planning and landscape architectural activities
		71.12	Engineering activities and related technical consultancy
		71.12/1	Engineering design activities for industrial process and production
		71.12/2	Engineering related scientific and technical consulting activities
		71.12/9	Other engineering activities (not including engineering design for industrial process and production or engineering related scientific and technical consulting activities)
	71.2		Technical testing and analysis
		71.20	Technical testing and analysis
72			**Scientific research and development**
	72.1		Research and experimental development on natural sciences and engineering
		72.11	Research and experimental development on biotechnology
		72.19	Other research and experimental development on natural sciences and engineering
	72.2		Research and experimental development on social sciences and humanities
		72.20	Research and experimental development on social sciences and humanities
73			**Advertising and market research**
	73.1		Advertising
		73.11	Advertising agencies
		73.12	Media representation
	73.2		Market research and public opinion polling
		73.20	Market research and public opinion polling
74			**Other professional, scientific and technical activities**
	74.1		Specialised design activities
		74.10	Specialised design activities
	74.2		Photographic activities
		74.20	Photographic activities
		74.20/1	Portrait photographic activities
		74.20/2	Other specialist photography (not including portrait photography)
		74.20/3	Film processing
		74.20/9	Other photographic activities (not including portrait and other specialist photography and film processing) n.e.c.
	74.3		Translation and interpretation activities
		74.30	Translation and interpretation activities
	74.9		Other professional, scientific and technical activities n.e.c.
		74.90	Other professional, scientific and technical activities n.e.c.
		74.90/1	Environmental consulting activities
		74.90/2	Quantity surveying activities
		74.90/9	Other professional, scientific and technical activities (not including environmental consultancy or quantity surveying) n.e.c.

M

Division	Group	Class and Subclass	Description
75			**Veterinary activities**
	75.0		Veterinary activities
		75.00	Veterinary activities
Section N			**Administrative and Support Service Activities**
77			**Rental and leasing activities**
	77.1		Renting and leasing of motor vehicles
		77.11	Renting and leasing of cars and light motor vehicles
		77.12	Renting and leasing of trucks
	77.2		Renting and leasing of personal and household goods
		77.21	Renting and leasing of recreational and sports goods
		77.22	Renting of video tapes and disks
		77.29	Renting and leasing of other personal and household goods
		77.29/1	Renting and leasing of media entertainment equipment
		77.29/9	Renting and leasing of other personal and household goods (other than media entertainment equipment)
	77.3		Renting and leasing of other machinery, equipment and tangible goods
		77.31	Renting and leasing of agricultural machinery and equipment
		77.32	Renting and leasing of construction and civil engineering machinery and equipment
		77.33	Renting and leasing of office machinery and equipment (including computers)
		77.34	Renting and leasing of water transport equipment
		77.34/1	Renting and leasing of passenger water transport equipment
		77.34/2	Renting and leasing of freight water transport equipment
		77.35	Renting and leasing of air transport equipment
		77.35/1	Renting and leasing of passenger air transport equipment
		77.35/2	Renting and leasing of freight air transport equipment
		77.39	Renting and leasing of other machinery, equipment and tangible goods n.e.c.
	77.4		Leasing of intellectual property and similar products, except copyrighted works
		77.40	Leasing of intellectual property and similar products, except copyrighted works
78			**Employment activities**
	78.1		Activities of employment placement agencies
		78.10	Activities of employment placement agencies
		78.10/1	Motion picture, television and other theatrical casting
		78.10/9	Activities of employment placement agencies (other than motion picture, television and other theatrical casting) n.e.c.
	78.2		Temporary employment agency activities
		78.20	Temporary employment agency activities
	78.3		Other human resources provision
		78.30	Other human resources provision
79			**Travel agency, tour operator and other reservation service and related activities**
	79.1		Travel agency and tour operator activities
		79.11	Travel agency activities
		79.12	Tour operator activities
	79.9		Other reservation service and related activities
		79.90	Other reservation service and related activities
		79.90/1	Activities of tourist guides
		79.90/9	Other reservation service and related activities (not including activities of tourist guides)

M
N

Division	Group	Class and Subclass	Description
80			**Security and investigation activities**
	80.1		Private security activities
		80.10	Private security activities
	80.2		Security systems service activities
		80.20	Security systems service activities
	80.3		Investigation activities
		80.30	Investigation activities
81			**Services to buildings and landscape activities**
	81.1		Combined facilities support activities
		81.10	Combined facilities support activities
	81.2		Cleaning activities
		81.21	General cleaning of buildings
		81.22	Other building and industrial cleaning activities
		81.22/1	Window cleaning services
		81.22/2	Specialised cleaning services
		81.22/3	Furnace and chimney cleaning services
		81.22/9	Building and industrial cleaning activities (other than window cleaning, specialised cleaning and furnace and chimney cleaning services) n.e.c.
		81.29	Other cleaning activities
		81.29/1	Disinfecting and extermination services
		81.29/9	Cleaning services (other than disinfecting and extermination services) n.e.c.
	81.3		Landscape service activities
		81.30	Landscape service activities
82			**Office administrative, office support and other business support activities**
	82.1		Office administrative and support activities
		82.11	Combined office administrative service activities
		82.19	Photocopying, document preparation and other specialised office support activities
	82.2		Activities of call centres
		82.20	Activities of call centres
	82.3		Organisation of conventions and trade shows
		82.30	Organisation of conventions and trade shows
		82.30/1	Activities of exhibition and fair organisers
		82.30/2	Activities of conference organisers
	82.9		Business support service activities n.e.c.
		82.91	Activities of collection agencies and credit bureaus
		82.91/1	Activities of collection agencies
		82.91/2	Activities of credit bureaus
		82.92	Packaging activities
		82.99	Other business support service activities n.e.c.
Section O			**Public Administration and Defence; Compulsory Social Security**
84			**Public administration and defence; compulsory social security**
	84.1		Administration of the State and the economic and social policy of the community
		84.11	General public administration activities

Division	Group	Class and Subclass	Description
		84.12	Regulation of the activities of providing health care, education, cultural services and other social services, excluding social security
		84.13	Regulation of and contribution to more efficient operation of businesses
	84.2		Provision of services to the community as a whole
		84.21	Foreign affairs
		84.22	Defence activities
		84.23	Justice and judicial activities
		84.24	Public order and safety activities
		84.25	Fire service activities
	84.3		Compulsory social security activities
		84.30	Compulsory social security activities
Section P			**Education**
85			**Education**
	85.1		Pre-primary education
		85.10	Pre-primary education
	85.2		Primary education
		85.20	Primary education
	85.3		Secondary education
		85.31	General secondary education
		85.32	Technical and vocational secondary education
	85.4		Higher education
		85.41	Post-secondary non-tertiary education
		85.42	Tertiary education
		85.42/1	First-degree level higher education
		85.42/2	Post-graduate level higher education
	85.5		Other education
		85.51	Sports and recreation education
		85.52	Cultural education
		85.53	Driving school activities
		85.59	Other education n.e.c.
	85.6		Educational support activities
		85.60	Educational support activities
Section Q			**Human Health and Social Work Activities**
86			**Human health activities**
	86.1		Hospital activities
		86.10	Hospital activities
		86.10/1	Hospital activities
		86.10/2	Medical nursing home activities
	86.2		Medical and dental practice activities
		86.21	General medical practice activities
		86.22	Specialist medical practice activities
		86.23	Dental practice activities
	86.9		Other human health activities
		86.90	Other human health activities

O
P
Q

Division	Group	Class and Subclass	Description
87			**Residential care activities**
	87.1		Residential nursing care activities
		87.10	Residential nursing care activities
	87.2		Residential care activities for learning disabilities, mental health and substance abuse
		87.20	Residential care activities for learning disabilities, mental health and substance abuse
	87.3		Residential care activities for the elderly and disabled
		87.30	Residential care activities for the elderly and disabled
	87.9		Other residential care activities
		87.90	Other residential care activities
88			**Social work activities without accommodation**
	88.1		Social work activities without accommodation for the elderly and disabled
		88.10	Social work activities without accommodation for the elderly and disabled
	88.9		Other social work activities without accommodation
		88.91	Child day-care activities
		88.99	Other social work activities without accommodation n.e.c.
Section R			**Arts, Entertainment and Recreation**
90			**Creative, arts and entertainment activities**
	90.0		Creative, arts and entertainment activities
		90.01	Performing arts
		90.02	Support activities to performing arts
		90.03	Artistic creation
		90.04	Operation of arts facilities
91			**Libraries, archives, museums and other cultural activities**
	91.0		Libraries, archives, museums and other cultural activities
		91.01	Library and archive activities
		91.01/1	Library activities
		91.01/2	Archive activities
		91.02	Museum activities
		91.03	Operation of historical sites and buildings and similar visitor attractions
		91.04	Botanical and zoological gardens and nature reserve activities
92			**Gambling and betting activities**
	92.0		Gambling and betting activities
		92.00	Gambling and betting activities
93			**Sports activities and amusement and recreation activities**
	93.1		Sports activities
		93.11	Operation of sports facilities
		93.12	Activities of sport clubs
		93.13	Fitness facilities
		93.19	Other sports activities
		93.19/1	Activities of racehorse owners
		93.19/9	Other sports activities (not including activities of racehorse owners) n.e.c.
	93.2		Amusement and recreation activities
		93.21	Activities of amusement parks and theme parks
		93.29	Other amusement and recreation activities

Q
R

Division	Group	Class and Subclass	Description
Section S			**Other Service Activities**
94			**Activities of membership organisations**
	94.1		Activities of business, employers and professional membership organisations
		94.11	Activities of business and employers membership organisations
		94.12	Activities of professional membership organisations
	94.2		Activities of trade unions
		94.20	Activities of trade unions
	94.9		Activities of other membership organisations
		94.91	Activities of religious organisations
		94.92	Activities of political organisations
		94.99	Activities of other membership organisations n.e.c.
95			**Repair of computers and personal and household goods**
	95.1		Repair of computers and communication equipment
		95.11	Repair of computers and peripheral equipment
		95.12	Repair of communication equipment
	95.2		Repair of personal and household goods
		95.21	Repair of consumer electronics
		95.22	Repair of household appliances and home and garden equipment
		95.23	Repair of footwear and leather goods
		95.24	Repair of furniture and home furnishings
		95.25	Repair of watches, clocks and jewellery
		95.29	Repair of other personal and household goods
96			**Other personal service activities**
	96.0		Other personal service activities
		96.01	Washing and (dry-)cleaning of textile and fur products
		96.02	Hairdressing and other beauty treatment
		96.03	Funeral and related activities
		96.04	Physical well-being activities
		96.09	Other personal service activities n.e.c.
Section T			**Activities of Households as Employers; Undifferentiated Goods-and Services-Producing Activities of Households for Own Use**
97			**Activities of households as employers of domestic personnel**
	97.0		Activities of households as employers of domestic personnel
		97.00	Activities of households as employers of domestic personnel
98			**Undifferentiated goods- and services-producing activities of private households for own use**
	98.1		Undifferentiated goods-producing activities of private households for own use
		98.10	Undifferentiated goods-producing activities of private households for own use
	98.2		Undifferentiated service-producing activities of private households for own use
		98.20	Undifferentiated service-producing activities of private households for own use
Section U			**Activities of Extraterritorial Organisations and Bodies**
99			**Activities of extraterritorial organisations and bodies**
	99.0		Activities of extraterritorial organisations and bodies
		99.00	Activities of extraterritorial organisations and bodies

S T U

Section A   **Agriculture, Forestry and Fishing**

This section includes the exploitation of vegetable and animal natural resources, comprising the activities of growing crops, raising and breeding animals, harvesting timber and other plants, animals or animal products from a farm or their natural habitats.

01   **Crop and animal production, hunting and related service activities**

This division includes two basic activities, namely the production of crop products and production of animal products, covering also the forms of organic agriculture, the growing of genetically modified crops and the raising of genetically modified animals. This division includes growing of crops in open fields as well in greenhouses.

This division also includes service activities incidental to agriculture, as well as hunting, trapping and related activities.

Group 01.5 (Mixed farming) breaks with the usual principles for identifying main activity. It accepts that many agricultural holdings have reasonably balanced crop and animal production, and that it would be arbitrary to classify them in one category or the other.

Agricultural activity excludes any subsequent processing of the agricultural products (classified under divisions 10 and 11 (Manufacture of food products and beverages) and division 12 (Manufacture of tobacco products)), beyond that needed to prepare them for the primary markets. The preparation of products for the primary markets is included here.

The division excludes field construction (e.g. agricultural land terracing, drainage, preparing rice paddies etc.) classified in section F (Construction) and buyers and cooperative associations engaged in the marketing of farm products classified in section G. Also excluded is landscape care and maintenance, which is classified in class 81.30.

01.1   **Growing of non-perennial crops**

This group includes the growing of non-perennial crops, i.e. plants that do not last for more than two growing seasons. Included is the growing of these plants for the purpose of seed production.

01.11   **Growing of cereals (except rice), leguminous crops and oil seeds**

This class includes all forms of growing of cereals, leguminous crops and oil seeds in open fields. The growing of these crops is often combined within agricultural units.

This class includes:

- growing of cereals such as:
  - wheat
  - grain maize
  - sorghum
  - barley
  - rye
  - oats
  - millets
  - other cereals n.e.c.
- growing of leguminous crops such as:
  - beans
  - broad beans
  - chick peas
  - cow peas
  - lentils
  - lupines
  - peas
  - pigeon peas
  - other leguminous crops
- growing of oil seeds such as:
  - soya beans
  - groundnuts
  - cottonseed
  - castor bean

A

- linseed
- mustard seed
- niger seed
- rapeseed
- safflower seed
- sesame seed
- sunflower seed
- other oil seeds

This class excludes:

- *growing of rice, see 01.12*
- *growing of sweet corn, see 01.13*
- *growing of maize for fodder, see 01.19*
- *growing of oleaginous fruits, see 01.26*

01.12   **Growing of rice**

01.13   **Growing of vegetables and melons, roots and tubers**

This class includes:

- growing of leafy or stem vegetables such as:
  - artichokes
  - asparagus
  - cabbages
  - cauliflower and broccoli
  - lettuce and chicory
  - spinach
  - other leafy or stem vegetables
- growing of fruit bearing vegetables such as:
  - cucumbers and gherkins
  - eggplants (aubergines)
  - tomatoes
  - watermelons
  - cantaloupes
  - other melons and fruit-bearing vegetables
- growing of root, bulb or tuberous vegetables such as:
  - carrots
  - turnips
  - garlic
  - onions (incl. shallots)
  - leeks and other alliaceous vegetables
  - other root, bulb or tuberous vegetables
- growing of mushrooms and truffles
- growing of vegetable seeds, including sugar beet seeds, excluding other beet seeds
- growing of sugar beet
- growing of other vegetables
- growing of roots and tubers such as:
  - potatoes
  - sweet potatoes
  - cassava
  - yams
  - other roots and tubers

This class excludes:

- *growing of chillies, peppers (capsicum sop.) and other spices and aromatic crops, see 01.28*
- *growing of mushroom spawn, see 01.30*

01.14   **Growing of sugar cane**

This class excludes:

- *growing of sugar beet, see 01.13*

01.15　　　**Growing of tobacco**

This class excludes:

– *manufacture of tobacco products, see 12.00*

01.16　　　**Growing of fibre crops**

This class includes:

– growing of cotton
– growing of jute, kenaf and other textile bast fibres
– growing of flax and true hemp
– growing of sisal and other textile fibre of the genus agave
– growing of abaca, ramie and other vegetable textile fibres
– growing of other fibre crops

01.19　　　**Growing of other non-perennial crops**

This class includes the growing of all other non-perennial crops:

– growing of mangolds, fodder roots, clover, alfalfa, sainfoin, fodder maize and other grasses, forage kale and similar forage products
– growing of beet seeds (excluding sugar beet seeds) and seeds of forage plants
– growing of flowers
– production of cut flowers and flower buds
– growing of flower seeds

This class excludes:

– *growing of non-perennial spices, aromatic, drug and pharmaceutical crops, see 01.28*

01.2　　　**Growing of perennial crops**

This group includes the growing of perennial crops, i.e. plants that last for more than two growing seasons, either dying back after each season or growing continuously. Included is the growing of these plants for the purpose of seed production.

01.21　　　**Growing of grapes**

This class includes:

– growing of wine grapes and table grapes in vineyards

This class excludes:

– *manufacture of wine, see 11.02*

01.22　　　**Growing of tropical and subtropical fruits**

This class includes:

– growing of tropical and subtropical fruits:
  - avocados
  - bananas and plantains
  - dates
  - figs
  - mangoes
  - papayas
  - pineapples
  - other tropical and subtropical fruits

01.23　　　**Growing of citrus fruits**

This class includes:

– growing of citrus fruits:
  - grapefruit and pomelo
  - lemons and limes
  - oranges
  - tangerines, mandarins and clementines
  - other citrus fruits

01.24　　　**Growing of pome fruits and stone fruits**

This class includes:

– growing of pome fruits and stone fruits:

A

- apples
- apricots
- cherries and sour cherries
- peaches and nectarines
- pears and quinces
- plums and sloes
- other pome fruits and stone fruits

01.25   **Growing of other tree and bush fruits and nuts**

This class includes:

- growing of berries:
  - blueberries
  - currants
  - gooseberries
  - kiwi fruit
  - raspberries
  - strawberries
  - other berries
- growing of fruit seeds
- growing of edible nuts:
  - almonds
  - cashew nuts
  - chestnuts
  - hazelnuts
  - pistachios
  - walnuts
  - other nuts
- growing of other tree and bush fruits
  - locust beans

This class excludes:

- *growing of coconuts, see 01.26*

01.26   **Growing of oleaginous fruits**

This class includes:

- growing of oleaginous fruits:
  - coconuts
  - olives
  - oil palms
  - other oleaginous fruits

This class excludes:

- *growing of soya beans, groundnuts and other oil seeds, see 01.11*

01.27   **Growing of beverage crops**

This class includes:

- growing of beverage crops:
  - coffee
  - tea
  - maté
  - cocoa
  - other beverage crops

01.28   **Growing of spices, aromatic, drug and pharmaceutical crops**

This class includes:

- growing of perennial and non-perennial spices and aromatic crops:
  - pepper (piper sop.)
  - chillies and peppers (capsicum sop.)
  - nutmeg, mace and cardamoms

- anise, badian and fennel
- cinnamon (canella)
- cloves
- ginger
- vanilla
- hops
- other spices and aromatic crops
- growing of drug and narcotic crops

01.29   Growing of other perennial crops

This class includes:

- growing of rubber trees for harvesting of latex
- growing of Christmas trees
- growing of trees for extraction of sap
- growing of vegetable materials of a kind used primarily for plaiting

This class excludes:

- *growing of flowers, production of cut flower buds and growing of flower seeds, see 01.19*
- *gathering of tree sap or rubber-like gums in the wild, see 02.30*

01.3   Plant propagation

01.30   Plant propagation

This class includes the production of all vegetative planting materials including cuttings, suckers and seedlings for direct plant propagation or to create plant grafting stock into which selected scion is grafted for eventual planting to produce crops.

This class includes:

- growing of plants for planting
- growing of plants for ornamental purposes, including turf for transplanting
- growing of live plants for bulbs, tubers and roots; cuttings and slips; mushroom spawn
- operation of tree nurseries, except forest tree nurseries

This class excludes:

- *growing of plants for the purpose of seed production, see 01.1, 01.2*
- *operation of forest tree nurseries, see 02.10*

01.4   Animal production

This group includes raising (farming) and breeding of all animals, except aquatic animals.

This class excludes:

- *farm animal boarding and care, see 01.62*
- *production of hides and skins from slaughterhouses, see 10.11*

01.41   Raising of dairy cattle

This class includes:

- raising and breeding of dairy cattle
- production of raw milk from cows or buffaloes

This class excludes:

- *processing of milk, see 10.51*

01.42   Raising of other cattle and buffaloes

This class includes:

- raising and breeding of cattle and buffaloes for meat
- production of bovine semen

01.43   Raising of horses and other equines

This class includes:

- raising and breeding of horses, asses, mules or hinnies

This class excludes:

- *operation of racing and riding stables, see 93.19*

A

01.44   **Raising of camels and camelids**

This class includes:

– raising and breeding of camels (dromedary) and camelids

01.45   **Raising of sheep and goats**

This class includes:

– raising and breeding of sheep and goats
– production of raw sheep or goat milk
– production of raw wool

This class excludes:

– *sheep shearing on a fee or contract basis, see 01.62*
– *production of pulled wool, see 10.11*
– *processing of milk, see 10.51*

01.46   **Raising of swine/pigs**

01.47   **Raising of poultry**

This class includes:

– raising and breeding of poultry:
  ■ chickens, turkeys, ducks, geese and guinea fowls
– production of eggs from poultry
– operation of poultry hatcheries

This class excludes:

– *production of feathers or down, see 10.12*

01.49   **Raising of other animals**

This class includes:

– raising and breeding of semi-domesticated or other live animals:
  ■ ostriches and emus
  ■ other birds (except poultry)
  ■ insects
  ■ rabbits and other fur animals
– production of fur skins, reptile or bird skins from ranching operation
– operation of worm farms, land mollusc farms, snail farms etc.
– raising of silk worms, production of silk worm cocoons
– bee-keeping and production of honey and beeswax
– raising and breeding of pet animals (except fish):
  ■ cats and dogs
  ■ birds, such as parakeets etc.
  ■ hamsters etc.
– raising of diverse animals

This class excludes:

– *production of hides and skins originating from hunting and trapping, see 01.70*
– *operation of frog farms, crocodile farms, marine worm farms, see 03.21, 03.22*
– *operation of fish farms, see 03.21, 03.22*
– *boarding and training of pet animals, see 96.09*
– *raising and breeding of poultry, see 01.47*

01.5   **Mixed farming**

01.50   Mixed farming

This class includes the combined production of crops and animals without a specialised production of crops or animals. The size of the overall farming operation is not a determining factor. If either production of crops or animals in a given unit is 66 per cent or more of standard gross margins, the combined activity should not be included here, but allocated to crop or animal farming.

This class excludes:

– *mixed crop farming, see groups 01.1 and 01.2*
– *mixed animal farming, see group 01.4*

A

01.6   **Support activities to agriculture and post-harvest crop activities**

This group includes activities incidental to agricultural production and activities similar to agriculture not undertaken for production purposes (in the sense of harvesting agricultural products), done on a fee or contract basis. Also included are post-harvest crop activities, aimed at preparing agricultural products for the primary market.

01.61   **Support activities for crop production**

This class includes:

– agricultural activities on a fee or contract basis:
  ▪ preparation of fields
  ▪ establishing a crop
  ▪ treatment of crops
  ▪ crop spraying, including by air
  ▪ trimming of fruit trees and vines
  ▪ transplanting of rice, thinning of beets
  ▪ harvesting
  ▪ pest control (including rabbits) in connection with agriculture
– maintenance of agricultural land in good agricultural and environmental condition
– operation of agricultural irrigation equipment

This class also includes:

– provision of agricultural machinery with operators and crew

This class excludes:

– *post-harvest crop activities, see 01.63*
– *drainage of agricultural land, see 43.12*
– *landscape architecture, see 71.11*
– *activities of agronomists and agricultural economists, see 74.90*
– *landscape gardening, planting, see 81.30*
– *organisation of agricultural shows and fairs, see 82.30*

01.62   **Support activities for animal production**

01.62/1   **Farm animal boarding and care**

This subclass includes:

– farm animal boarding and care

This subclass excludes:

– *provision of space for animal boarding only, see 68.20*
– *pet boarding, see 96.09*

01.62/9   **Support activities for animal production (other than farm animal boarding and care) n.e.c.**

This subclass includes:

– agricultural activities on a fee or contract basis:
  ▪ activities to promote propagation, growth and output of animals
  ▪ herd testing services, droving services, agistment services, poultry caponising, coop cleaning etc.
  ▪ activities related to artificial insemination
  ▪ stud services
  ▪ sheep shearing

This subclass also includes:

– activities of farriers

This subclass excludes:

– *veterinary activities, see 75.00*
– *vaccination of animals, see 75.00*
– *renting of animals (e.g. herds), see 77.39*

01.63   **Post-harvest crop activities**

This class includes:

– preparation of crops for primary markets, i.e. cleaning, trimming, grading, disinfecting

A

- cotton ginning
- preparation of tobacco leaves, e.g. drying
- preparation of cocoa beans, e.g. peeling
- waxing of fruit
- sun-drying of fruit and vegetables

This class excludes:

- *preparation of agricultural products by the producer, see corresponding class in groups 01.1, 01.2 or 01.3*
- *post-harvest activities aimed at improving the propagation quality of seed, see 01.64*
- *stemming and redrying of tobacco, see 12.00*
- *marketing activities of commission merchants and cooperative associations, see division 46*
- *wholesale of agricultural raw materials, see 46.2*

01.64   Seed processing for propagation

This class includes all post-harvest activities aimed at improving the propagation quality of seed through the removal of non-seed materials, undersized, mechanically or insect-damaged and immature seeds as well as removing the seed moisture to a safe level for seed storage. This activity includes the drying, cleaning, grading and treating of seeds until they are marketed. The treatment of genetically modified seeds is included here.

This class excludes:

- *growing of seeds, see groups 01.1 and 01.2*
- *processing of seeds to obtain oil, see 10.41*
- *research to develop or modify new forms of seeds, see 72.11*

01.7   **Hunting, trapping and related service activities**

01.70   Hunting, trapping and related service activities

This class includes:

- hunting and trapping on a commercial basis
- taking of animals (dead or alive) for food, fur, skin, or for use in research, in zoos or as pets
- production of fur skins, reptile or bird skins from hunting or trapping activities

This class also includes:

- land-based catching of sea mammals such as walrus and seal

This class excludes:

- *production of fur skins, reptile or bird skins from ranching operations, see group 01.49*
- *raising of game animals on ranching operations, see 01.4*
- *catching of whales, see 03.11*
- *production of hides and skins originating from slaughterhouses, see 10.11*
- *hunting for sport or recreation and related service activities, see 93.19*
- *service activities to promote hunting and trapping, see 94.99*

02   **Forestry and logging**

This division includes the production of roundwood as well as the extraction and gathering of wild growing non-wood forest products. Besides the production of timber, forestry activities result in products that undergo little processing, such as firewood, charcoal and roundwood used in an unprocessed form (e.g. pit-props, pulpwood etc.). These activities can be carried out in natural or planted forests.

Excluded is further processing of wood beginning with sawmilling and planning of wood, see division 16.

02.1   **Silviculture and other forestry activities**

02.10   Silviculture and other forestry activities

This class includes:

- growing of standing timber: planting, replanting, transplanting, thinning and conserving of forests and timber tracts
- growing of coppice, pulpwood and fire wood
- operation of forest tree nurseries

These activities can be carried out in natural or planted forests.

This class excludes:

- *growing of Christmas trees, see 01.29*

A

– *operation of tree nurseries, except for forest trees, see 01.30*
– *gathering of mushrooms and other wild growing non-wood forest products, see 02.30*
– *production of wood chips and particles, see 16.10*

02.2   Logging

02.20   Logging

This class includes:

– production of roundwood for forest-based manufacturing industries
– production of roundwood used in an unprocessed form such as pit-props, fence posts and utility poles
– gathering and production of wood for energy
– gathering and production of forest harvesting residues for energy
– production of charcoal in the forest (using traditional methods)

The output of this activity can take the form of logs or fire wood.

This class excludes:

– *growing of Christmas trees, see 01.29*
– *growing of standing timber: planting, replanting, transplanting, thinning and conserving of forests and timber tracts, see 02.10*
– *gathering of wild growing non-wood forest products, see 02.30*
– *production of wood chips and particles, see 16.10*
– *production of charcoal through distillation of wood, see 20.14*

02.3   Gathering of wild growing non-wood products

02.30   Gathering of wild growing non-wood products

This class includes:

– gathering of wild growing materials:
  ■ mushrooms, truffles
  ■ berries
  ■ nuts
  ■ balata and other rubber-like gums
  ■ cork
  ■ lac and resins
  ■ balsams
  ■ vegetable hair
  ■ eelgrass
  ■ acorns, horse chestnuts
  ■ mosses and lichens

This class excludes:

– *managed production of any of these products (except growing of cork trees), see division 01*
– *growing of mushrooms or truffles, see 01.13*
– *growing of berries or nuts, see 01.25*
– *gathering of fire wood, see 02.20*
– *production of wood chips, see 16.10*

02.4   Support services to forestry

02.40   Support services to forestry

This class includes carrying out part of the forestry operation on a fee or contract basis.

This class includes:

– forestry service activities:
  ■ forestry inventories
  ■ forest management consulting services
  ■ timber evaluation
  ■ forest fire fighting and protection
  ■ forest pest control
– logging service activities:
  ■ transport of logs within the forest

A

This class excludes:

– *operation of forest tree nurseries, see 02.10*
– *draining of forestry land, see 43.12*
– *clearing of building sites, see 43.12*

03   Fishing and aquaculture

This division includes capture fishery and aquaculture, covering the use of fishery resources from marine, brackish or freshwater environments, with the goal of capturing or gathering fish, crustaceans, molluscs and other marine organisms and products (e.g. aquatic plants, pearls, sponges etc). Also included are activities that are normally integrated in the process of production for own account (e.g. seeding oysters for pearl production). Service activities incidental to marine or freshwater fishery or aquaculture are included in the related fishing or aquaculture activities.

This division does not include building and repairing of ships and boats (30.1, 33.15) and sport or recreational fishing activities (93.19). Processing of fish, crustaceans or molluscs is excluded, whether at land-based plants or on factory ships (10.20).

03.1   Fishing

This group includes "capture fishery", i.e. the hunting, collecting and gathering activities directed at removing or collecting live wild aquatic organisms (predominantly fish, molluscs and crustaceans) including plants from the oceanic, coastal or inland waters for human consumption and other purposes by hand or more usually by various types of fishing gear such as nets, lines and stationary traps. Such activities can be conducted on the intertidal shoreline (e.g. collection of molluscs such as mussels and oysters) or shore based netting, or from home-made dugouts or more commonly using commercially made boats in inshore, coastal waters or offshore waters. Such activities also include fishing in restocked water bodies.

03.11   Marine fishing

This class includes:

– fishing on a commercial basis in ocean and coastal waters
– taking of marine crustaceans and molluscs
– whale catching
– taking of marine aquatic animals: turtles, sea squirts, tunicates, sea urchins etc.

This class also includes:

– activities of vessels engaged both in marine fishing, and in processing and preserving of fish
– gathering of other marine organisms and materials: natural pearls, sponges, coral and algae

This class excludes:

– *capturing of marine mammals, except whales, e.g. walruses, seals, see 01.70*
– *processing of whales on factory ships, see 10.11*
– *processing of fish, crustaceans and molluscs on factory ships or in factories ashore, see 10.20*
– *renting of pleasure boats with crew for sea and coastal water transport (e.g. for fishing cruises), see 50.10*
– *fishing inspection, protection and patrol services, see 84.24*
– *fishing practiced for sport or recreation and related services, see 93.19*
– *operation of sport fishing preserves, see 93.19*

03.12   Freshwater fishing

This class includes:

– fishing on a commercial basis in inland waters
– taking of freshwater crustaceans and molluscs
– taking of freshwater aquatic animals

This class also includes:

– gathering of freshwater materials

This class excludes:

– *processing of fish, crustaceans and molluscs, see 10.20*
– *fishing inspection, protection and patrol services, see 84.24*
– *fishing practiced for sport or recreation and related services, see 93.19*
– *operation of sport fishing preserves, see 93.19*

03.2   Aquaculture

This group includes "aquaculture" (or aquafarming), i.e. the production process involving the culturing or farming (including harvesting) of aquatic organisms (fish, molluscs, crustaceans, plants, crocodiles, alligators and amphibians) using techniques designed to increase the production of the organisms in question beyond the natural capacity of the environment (for example regular stocking, feeding and protection from predators).

Culturing/farming refers to the rearing up to their juvenile and/or adult phase under captive conditions of the above organisms.

In addition, "aquaculture" also encompasses individual, corporate or state ownership of the individual organisms throughout the rearing or culture stage, up to and including harvesting.

03.21　　**Marine aquaculture**

This class includes:

– fish farming in sea water including farming of marine ornamental fish
– production of bivalve spat (oyster mussel etc.), lobsterlings, shrimp post-larvae, fish fry and fingerlings
– growing of laver and other edible seaweeds
– culture of crustaceans, bivalves, other molluscs and other aquatic animals in sea water

This class also includes:

– aquaculture activities in brackish waters
– aquaculture activities in salt water filled tanks or reservoirs
– operation of fish hatcheries (marine)
– operation of marine worm farms

This class excludes:

– *frog farming, see 03.22*
– *operation of sport fishing preserves, see 93.19*

03.22　　**Freshwater aquaculture**

This class includes:

– fish farming in freshwater including farming of freshwater ornamental fish
– culture of freshwater crustaceans, bivalves, other molluscs and other aquatic animals
– operation of fish hatcheries (freshwater)
– farming of frogs

This class excludes:

– *aquaculture activities in salt water filled tanks and reservoirs, see 03.21*
– *operation of sport fishing preserves, see 93.19*

B

Section B   **Mining and Quarrying**

Mining and quarrying include the extraction of minerals occurring naturally as solids (coal and ores), liquids (petroleum) or gases (natural gas). Extraction can be achieved by different methods such as underground or surface mining, well operation, seabed mining etc.

This section includes supplementary activities aimed at preparing the crude materials for marketing, for example, crushing, grinding, cleaning, drying, sorting, concentrating ores, liquefaction of natural gas and agglomeration of solid fuels. These operations are often accomplished by the units that extracted the resource and/or others located nearby.

Mining activities are classified into divisions, groups and classes on the basis of the principal mineral produced. Divisions 05, 06 are concerned with mining and quarrying of fossil fuels (coal, lignite, petroleum, gas); divisions 07, 08 concern metal ores, various minerals and quarry products.

Some of the technical operations of this section, particularly related to the extraction of hydrocarbons, may also be carried out for third parties by specialised units as an industrial service which is reflected in division 09.

This section excludes:

- *processing of the extracted materials, see section C (Manufacturing)*
- *use of the extracted materials without a further transformation for construction purposes, see section F (Construction)*
- *bottling of natural spring and mineral waters at springs and wells, see 11.07*
- *crushing, grinding or otherwise treating certain earths, rocks and minerals not carried on in conjunction with mining and quarrying, see 23.9*

05   **Mining of coal and lignite**

This division includes the extraction of solid mineral fuels through underground or open-cast mining and includes operations (e.g. grading, cleaning, compressing and other steps necessary for transportation etc.) leading to a marketable product.

This division does not include coking (see 19.10), services incidental to coal or lignite mining (see 09.90) or the manufacture of briquettes (see 19.20).

05.1   **Mining of hard coal**

05.10   Mining of hard coal

05.10/1   Mining of hard coal from deep coal mines (underground mining)

This subclass includes:

- mining of hard coal: underground mining, including mining through liquefaction methods
- cleaning, sizing, grading, pulverising, compressing etc. of coal from underground mining to classify, improve quality or facilitate transport or storage

This subclass excludes:

- *lignite mining, see 05.20*
- *peat digging, see 08.92*
- *support activities for hard coal mining, see 09.90*
- *test drilling for coal mining, see 09.90*
- *coke ovens producing solid fuels, see 19.10*
- *manufacture of hard coal briquettes, see 19.20*
- *work performed to develop or prepare properties for coal mining, see 43.12*

05.10/2   Mining of hard coal from open cast coal working (surface mining)

This subclass includes:

- mining of hard coal: surface mining
- cleaning, sizing, grading, pulverising, compressing etc. of coal from surface mining to classify, improve quality or facilitate transport or storage

This subclass also includes:

- recovery of hard coal from culm banks

This subclass excludes:

- *lignite mining, see 05.20*
- *peat digging, see 08.92*
- *support activities for hard coal mining, see 09.90*
- *test drilling for coal mining, see 09.90*
- *coke ovens producing solid fuels, see 19.10*

- *manufacture of hard coal briquettes, see 19.20*
- *work performed to develop or prepare properties for coal mining, see 43.12*

05.2　Mining of lignite

05.20　Mining of lignite

This class includes:

- mining of lignite (brown coal): underground or surface mining, including mining through liquefaction methods
- washing, dehydrating, pulverising, compressing of lignite to improve quality or facilitate transport or storage

This class excludes:

- *hard coal mining, see 05.10*
- *peat digging, see 08.92*
- *support activities for lignite mining, see 09.90*
- *test drilling for coal mining, see 09.90*
- *manufacture of lignite fuel briquettes, see 19.20*
- *work performed to develop or prepare properties for coal mining, see 43.12*

06　Extraction of crude petroleum and natural gas

This division includes the production of crude petroleum, the mining and extraction of oil from oil shale and oil sands, and the production of natural gas and recovery of hydrocarbon liquids. This division includes the activities of operating and/or developing oil and gas field properties. Such activities may include drilling, completing, and equipping wells; operating separators, emulsion breakers, desalting equipment, and field gathering lines for crude petroleum; and all other activities in the preparation of oil and gas up to the point of shipment from the producing property.

This division excludes:

- *oil and gas field services, performed on a fee or contract basis, see 09.10*
- *oil and gas well exploration, see 09.10*
- *test drilling and boring, see 09.10*
- *refining of petroleum products, see 19.20*
- *geophysical, geologic and seismic surveying, see 71.12*

06.1　Extraction of crude petroleum

06.10　Extraction of crude petroleum

This class includes:

- extraction of crude petroleum oils

This class also includes:

- extraction of bituminous or oil shale and tar sand
- production of crude petroleum from bituminous shale and sand
- processes to obtain crude oils: decantation, desalting, dehydration, stabilisation etc.

This class excludes:

- *support activities for oil and natural gas extraction, see 09.10*
- *oil and gas exploration, see 09.10*
- *manufacture of refined petroleum products, see 19.20*
- *recovery of liquefied petroleum gases in the refining of petroleum, see 19.20*
- *operation of pipelines, see 49.50*

06.2　Extraction of natural gas

06.20　Extraction of natural gas

This class includes:

- production of crude gaseous hydrocarbon (natural gas)
- extraction of condensates
- draining and separation of liquid hydrocarbon fractions
- gas desulphurisation

This class also includes:

- mining of hydrocarbon liquids, obtained through liquefaction or pyrolysis

This class excludes:

– *support activities for oil and natural gas extraction, see 09.10*
– *oil and gas exploration, see 09.10*
– *recovery of liquefied petroleum gases in the refining of petroleum, see 19.20*
– *manufacture of industrial gases, see 20.11*
– *operation of pipelines, see 49.50*

07   Mining of metal ores

This division includes mining for metallic minerals (ores), performed through underground or open-cast extraction, seabed mining etc.

Also included are ore dressing and beneficiating operations, such as crushing, grinding, washing, drying, sintering, calcining or leaching ore, gravity separation or flotation operations.

This division excludes:

– *roasting of iron pyrites, see 20.13*
– *production of aluminium oxide, see 24.42*
– *operation of blast furnaces, see division 24*

07.1   Mining of iron ores

07.10   Mining of iron ores

This class includes:

– mining of ores valued chiefly for iron content
– beneficiation and agglomeration of iron ores

This class excludes:

– *extraction and preparation of pyrites and pyrrhotite (except roasting), see 08.91*

07.2   Mining of non-ferrous metal ores

This group includes the mining of non-ferrous metal ores.

07.21   Mining of uranium and thorium ores

This class includes:

– mining of ores chiefly valued for uranium and thorium content: pitchblende etc.
– concentration of such ores
– manufacture of yellowcake

This class excludes:

– *enrichment of uranium and thorium ores, see 20.13*
– *production of uranium metal from pitchblende or other ores, see 24.46*
– *smelting and refining of uranium, see 24.46*

07.29   Mining of other non-ferrous metal ores

This class includes:

– mining and preparation of ores chiefly valued for non-ferrous metal content:
  ▪ aluminium (bauxite), copper, lead, zinc, tin, manganese, chrome, nickel, cobalt, molybdenum, tantalum, vanadium etc.
  ▪ precious metals: gold, silver, platinum

This class excludes:

– *mining and preparation of uranium and thorium ores, see 07.21*
– *production of aluminium oxide, see 24.42*
– *production of mattes of copper or of nickel, see 24.44, 24.45*

08   Other mining and quarrying

This division includes extraction from a mine or quarry, but also dredging of alluvial deposits, rock crushing and the use of salt marshes. The products are used most notably in construction (e.g. sands, stones etc.), manufacture of materials (e.g. clay, gypsum, calcium etc.), manufacture of chemicals etc.

This division does not include processing (except crushing, grinding, cutting, cleaning, drying, sorting and mixing) of the minerals extracted.

B

08.1   **Quarrying of stone, sand and clay**

08.11   Quarrying of ornamental and building stone, limestone, gypsum, chalk and slate

This class includes:

- quarrying, rough trimming and sawing of monumental and building stone such as marble, granite, sandstone etc.
- breaking and crushing of ornamental and building stone
- quarrying, crushing and breaking of limestone
- mining of gypsum and anhydrite
- mining of chalk and uncalcined dolomite

This class excludes:

- *mining of chemical and fertiliser minerals, see 08.91*
- *production of calcined dolomite, see 23.52*
- *cutting, shaping and finishing of stone outside quarries, see 23.70*

08.12   Operation of gravel and sand pits; mining of clays and kaolin

This class includes:

- extraction and dredging of industrial sand, sand for construction and gravel
- breaking and crushing of gravel
- quarrying of sand
- mining of clays, refractory clays and kaolin

This class excludes:

- *mining of bituminous sand, see 06.10*

08.9   **Mining and quarrying n.e.c.**

08.91   Mining of chemical and fertiliser minerals

This class includes:

- mining of natural phosphates and natural potassium salts
- mining of native sulphur
- extraction and preparation of pyrites and pyrrhotite, except roasting
- mining of natural barium sulphate and carbonate (barytes and witherite), natural borates, natural magnesium sulphates (kieserite)
- mining of earth colours, fluorspar and other minerals valued chiefly as a source of chemicals

This class also includes:

- guano mining

This class excludes:

- *extraction of salt, see 08.93*
- *roasting of iron pyrites, see 20.13*
- *manufacture of synthetic fertilisers and nitrogen compounds, see 20.15*

08.92   Extraction of peat

This class includes:

- peat digging
- preparation of peat to improve quality or facilitate transport or storage

This class excludes:

- *service activities incidental to peat mining, see 09.90*
- *manufacturing of peat briquettes, see 19.20*
- *manufacture of potting soil mixtures of peat, natural soil, sands, clays, fertiliser minerals etc., see 20.15*
- *manufacture of articles of peat, see 23.99*

08.93   Extraction of salt

This class includes:

- extraction of salt from underground including by dissolving and pumping
- salt production by evaporation of sea water or other saline waters
- crushing, purification and refining of salt by the producer

B

This class excludes:

- *processing of salt into food-grade salt, e.g. iodised salt, see 10.84*
- *potable water production by evaporation of saline water, see 36.00*

08.99   **Other mining and quarrying n.e.c.**

This class includes:

- mining and quarrying of various minerals and materials:
  - abrasive materials, asbestos, siliceous fossil meals, natural graphite, steatite (talc), feldspar etc.
  - natural asphalt, asphaltites and asphaltic rock; natural solid bitumen
  - gemstones, quartz, mica etc.

09   **Mining support service activities**

This division includes specialised support services incidental to mining provided on a fee or contract basis. It includes exploration services through traditional prospecting methods such as taking core samples and making geological observations as well as drilling, test-drilling or redrilling for oil wells, metallic and non-metallic minerals. Other typical services cover building oil and gas well foundations, cementing oil and gas well casings, cleaning, bailing and swabbing oil and gas wells, draining and pumping mines, overburden removal services at mines, etc.

09.1   **Support activities for petroleum and natural gas extraction**

09.10   Support activities for petroleum and natural gas extraction

This class includes:

- oil and gas extraction service activities provided on a fee or contract basis:
  - exploration services in connection with petroleum or gas extraction, e.g. traditional prospecting methods, such as making geological observations at prospective sites
  - directional drilling and redrilling; "spudding in"; derrick erection in situ, repairing and dismantling; cementing oil and gas well casings; pumping of wells; plugging and abandoning wells etc.
  - liquefaction and regasification of natural gas for purpose of transport, done at the mine site
  - draining and pumping services, on a fee or contract basis
  - test drilling in connection with petroleum or gas extraction

This class also includes:

- oil and gas field fire fighting services

This class excludes:

- *service activities performed by operators of oil or gas fields, see 06.10, 06.20*
- *specialised repair of mining machinery, see 33.12*
- *liquefaction and regasification of natural gas for purpose of transport, done off the mine site, see 52.21*
- *geophysical, geologic and seismic surveying, see 71.12*

09.9   **Support activities for other mining and quarrying**

09.90   Support activities for other mining and quarrying

This class includes:

- support services on a fee or contract basis, required for mining activities of divisions 05, 07 and 08
  - exploration services, e.g. traditional prospecting methods, such as taking core samples and making geological observations at prospective sites
  - draining and pumping services, on a fee or contract basis
  - test drilling and test hole boring

This class excludes:

- *operating mines or quarries on a contract or fee basis, see division 05, 07 or 08*
- *specialised repair of mining machinery, see 33.12*
- *geophysical surveying services, on a contract or fee basis, see 71.12*

Section C   **Manufacturing**

This section includes the physical or chemical transformation of materials, substances, or components into new products, although this cannot be used as the single universal criterion for defining manufacturing (see remark on processing of waste below). The materials, substances, or components transformed are raw materials that are products of agriculture, forestry, fishing, mining or quarrying as well as products of other manufacturing activities. Substantial alteration, renovation or reconstruction of goods is generally considered to be manufacturing.

The output of a manufacturing process may be finished in the sense that it is ready for utilisation or consumption, or it may be semi-finished in the sense that it is to become an input for further manufacturing. For example, the output of alumina refining is the input used in the primary production of aluminium; primary aluminium is the input to aluminium wire drawing; and aluminium wire is the input for the manufacture of fabricated wire products.

Manufacture of specialised components and parts of, and accessories and attachments to, machinery and equipment is, as a general rule, classified in the same class as the manufacture of the machinery and equipment for which the parts and accessories are intended. Manufacture of unspecialised components and parts of machinery and equipment, e.g. engines, pistons, electric motors, electrical assemblies, valves, gears, roller bearings, is classified in the appropriate class of manufacturing, without regard to the machinery and equipment in which these items may be included. However, making specialised components and accessories by moulding or extruding plastics materials is included in group 22.2.

Assembly of the component parts of manufactured products is considered manufacturing. This includes the assembly of manufactured products from either self-produced or purchased components.

The recovery of waste, i.e. the processing of waste into secondary raw materials is classified in group 38.3 (Materials recovery). While this may involve physical or chemical transformations, this is not considered to be a part of manufacturing. The primary purpose of these activities is considered to be the treatment or processing of waste and they are therefore classified in Section E (Water supply; sewerage, waste management and remediation activities). However, the manufacture of new final products (as opposed to secondary raw materials) is classified in manufacturing, even if these processes use waste as an input. For example, the production of silver from film waste is considered to be a manufacturing process.

Specialised maintenance and repair of industrial, commercial and similar machinery and equipment is, in general, classified in division 33 (Repair, maintenance and installation of machinery and equipment). However, the repair of computers and personal and household goods is classified in division 95 (Repair of computers and personal and household goods), while the repair of motor vehicles is classified in division 45 (Wholesale and retail trade and repair of motor vehicles and motorcycles).

The installation of machinery and equipment, when carried out as a specialised activity, is classified in 33.20.

Remark: The boundaries of manufacturing and the other sectors of the classification system can be somewhat blurred. As a general rule, the activities in the manufacturing section involve the transformation of materials into new products. Their output is a new product. However, the definition of what constitutes a new product can be somewhat subjective. As clarification, the following activities are considered manufacturing in SIC:

- fresh fish processing (oyster shucking, fish filleting), not done on a fishing boat (see 10.20)
- milk pasteurising and bottling (see 10.51)
- leather converting (see 15.11)
- wood preserving (see 16.10)
- printing and related activities (see 18.1)
- tyre retreading (see 22.11)
- ready-mixed concrete production (see 23.63)
- electroplating, plating, and metal heat treating (see 25.61)
- rebuilding or remanufacturing of machinery (e.g. automobile engines, see 29.10)

Conversely, there are activities that, although sometimes involving transformation processes, are classified in other sections of SIC; in other words, they are not classified as manufacturing. They include:

- logging, classified in section A (Agriculture, forestry and fishing);
- beneficiating of agricultural products, classified in section A (Agriculture, forestry and fishing);
- preparation of food for immediate consumption on the premises is classified to division 56 (Food and beverage service activities)
- beneficiating of ores and other minerals, classified in section B (Mining and quarrying);
- construction of structures and fabricating operations performed at the site of construction, classified in section F (Construction);
- activities of breaking bulk and redistribution in smaller lots, including packaging, repackaging, or bottling products, such as liquors or chemicals; sorting of scrap; mixing paints to customer order; and cutting metals to customer order; treatment not resulting into a different good is classified to section G (Wholesale and retail trade; repair of motor vehicles and motorcycles).

10   **Manufacture of food products**

This division includes the processing of the products of agriculture, forestry and fishing into food for humans or animals, and includes the production of various intermediate products that are not directly food products. The activity often generates associated products of greater or lesser value (for example, hides from slaughtering, or oilcake from oil production).

This division is organised by activities dealing with different kinds of products: meat, fish, fruit and vegetables, fats and oils, milk products, grain mill products, animal feeds and other food products. Production can be carried out for own account, as well as for third parties, as in custom slaughtering.

Some activities are considered manufacturing (for example, those performed in bakeries, pastry shops, and prepared meat shops etc. which sell their own production) even though there is retail sale of the products in the producers' own shop. However, where the processing is minimal and does not lead to a real transformation, the unit is classified to wholesale and retail trade (section G).

Preparation of food for immediate consumption on the premises is classified to division 56 (Food and beverage service activities).

Production of animal feeds from slaughter waste or by-products is classified in 10.9, while processing food and beverage waste into secondary raw material is classified to 38.3, and disposal of food and beverage waste in 38.21.

This division does not include the preparation of meals for immediate consumption, such as in restaurants.

10.1   **Processing and preserving of meat and production of meat products**

10.11   Processing and preserving of meat

This class includes:

– operation of slaughterhouses engaged in killing, dressing or packing meat: beef, pork, lamb, rabbit, mutton, camel, etc.
– production of fresh, chilled or frozen meat, in carcasses
– production of fresh, chilled or frozen meat, in cuts

This class also includes:

– slaughtering and processing of whales on land or on specialised vessels
– production of hides and skins originating from slaughterhouses, including fellmongery
– rendering of lard and other edible fats of animal origin
– processing of animal offal
– production of pulled wool

This class excludes:

– *rendering of edible poultry fats, see 10.12*
– *packaging of meat, see 82.92*

10.12   Processing and preserving of poultry meat

This class includes:

– operation of slaughterhouses engaged in killing, dressing or packing poultry
– production of fresh, chilled or frozen meat in individual portions
– rendering of edible poultry fats

This class also includes:

– production of feathers and down

This class excludes:

– *packaging of meat, see 82.92*

10.13   Production of meat and poultry meat products

This class includes:

– production of dried, salted or smoked meat
– production of meat products:
  - sausages, salami, puddings, "andouillettes", saveloys, bolognas, pâtés, rillettes, boiled ham

This class excludes:

– *manufacture of prepared frozen meat and poultry dishes, see 10.85*
– *manufacture of soup containing meat, see 10.89*
– *wholesale trade of meat, see 46.32*
– *packaging of meat, see 82.92*

10.2　　　　**Processing and preserving of fish, crustaceans and molluscs**

10.20　　　Processing and preserving of fish, crustaceans and molluscs

This class includes:

– preparation and preservation of fish, crustaceans and molluscs: freezing, deep-freezing, drying, cooking, smoking, salting, immersing in brine, canning etc.
– production of fish, crustacean and mollusc products: fish fillets, roes, caviar, caviar substitutes etc.
– production of fishmeal for human consumption or animal feed
– production of meals and solubles from fish and other aquatic animals unfit for human consumption

This class also includes:

– activities of vessels engaged only in the processing and preserving of fish
– processing of seaweed

This class excludes:

– *processing and preserving of fish on vessels engaged in fishing, see 03.11*
– *processing of whales on land or specialised vessels, see 10.11*
– *production of oils and fats from marine material, see 10.41*
– *manufacture of prepared frozen fish dishes, see 10.85*
– *manufacture of fish soups, see 10.89*

10.3　　　　**Processing and preserving of fruit and vegetables**

10.31　　　Processing and preserving of potatoes

This class includes:

– processing and preserving of potatoes:
  ▪ manufacture of prepared frozen potatoes
  ▪ manufacture of dehydrated mashed potatoes
  ▪ manufacture of potato snacks
  ▪ manufacture of potato crisps
  ▪ manufacture of potato flour and meal

This class also includes:

– industrial peeling of potatoes

10.32　　　Manufacture of fruit and vegetable juice

This class includes:

– manufacture of fruit or vegetable juices

This class also includes:

– production of concentrates from fresh fruits and vegetables

10.39　　　Other processing and preserving of fruit and vegetables

This class includes:

– manufacture of food consisting chiefly of fruit or vegetables, except ready-made dishes in frozen or canned form
– preserving of fruit, nuts or vegetables: freezing, drying, immersing in oil or in vinegar, canning etc.
– manufacture of fruit or vegetable food products
– manufacture of jams, marmalades and table jellies
– roasting of nuts
– manufacture of nut foods and pastes

This class also includes:

– manufacture of perishable prepared foods of fruit and vegetables, such as:
  ▪ salads; mixed salads, packaged
  ▪ peeled or cut vegetables
  ▪ tofu (bean curd)

This class excludes:

– *manufacture of fruit or vegetable juices, see 10.32*
– *manufacture of flour or meal of dried leguminous vegetables, see 10.61*

      – *preservation of fruit and nuts in sugar, see 10.82*
      – *manufacture of prepared vegetable dishes, see 10.85*
      – *manufacture of artificial concentrates, see 10.89*

10.4   **Manufacture of vegetable and animal oils and fats**

This group includes the manufacture of crude and refined oils and fats from vegetable or animal materials, except rendering or refining of lard and other edible animal fats.

10.41   **Manufacture of oils and fats**

This class includes:

– manufacture of crude vegetable oils: olive oil, soya-bean oil, palm oil, sunflower-seed oil, cotton-seed oil, rape, colza or mustard oil, linseed oil etc.
– manufacture of non-defatted flour or meal of oilseeds, oil nuts or oil kernels
– manufacture of refined vegetable oils: olive oil, soya-bean oil etc.
– processing of vegetable oils: blowing, boiling, dehydration, hydrogenation etc.

This class also includes:

– manufacture of non-edible animal oils and fats
– extraction of fish and marine mammal oils
– production of cotton linters, oilcakes and other residual products of oil production

This class excludes:

– *rendering and refining of lard and other edible animal fats, see 10.11*
– *manufacture of margarine, see 10.42*
– *wet corn milling, see 10.62*
– *manufacture of corn oil, see 10.62*
– *production of essential oils, see 20.53*
– *treatment of oil and fats by chemical processes, see 20.59*

10.42   **Manufacture of margarine and similar edible fats**

This class includes:

– manufacture of margarine
– manufacture of melanges and similar spreads
– manufacture of compound cooking fats

10.5   **Manufacture of dairy products**

10.51   **Operation of dairies and cheese making**

10.51/1   **Liquid milk and cream production**

This subclass includes:

– manufacture of fresh liquid milk, pasteurised, sterilised, homogenised and/or ultra heat treated
– manufacture of cream from fresh liquid milk, pasteurised, sterilised, homogenised

This subclass excludes:

– *production of raw milk (cattle), see 01.41*
– *production of raw milk (sheep, goats, horses, asses, camels, etc.), see 01.43, 01.44, 01.45*
– *manufacture of non-dairy milk substitutes, see 10.89*

10.51/2   **Butter and cheese production**

This subclass includes:

– manufacture of butter
– manufacture of cheese and curd

This subclass excludes:

– *manufacture of non-dairy cheese substitutes, see 10.89*

10.51/9   **Manufacture of milk products (other than liquid milk and cream, butter, cheese) n.e.c.**

This subclass includes:

– manufacture of milk-based soft drinks
– manufacture of dried or concentrated milk whether or not sweetened

    – manufacture of milk or cream in solid form
    – manufacture of yoghurt
    – manufacture of whey
    – manufacture of casein or lactose

10.52   Manufacture of ice cream

This class includes:

– manufacture of ice cream and other edible ice such as sorbet

This class excludes:

– activities of ice cream parlours, see 56.10

10.6   Manufacture of grain mill products, starches and starch products

This group includes the milling of flour or meal from grains or vegetables, the milling, cleaning and polishing of rice, as well as the manufacture of flour mixes or doughs from these products. Also included in this group is the wet milling of corn and vegetables and the manufacture of starch and starch products.

10.61   Manufacture of grain mill products

10.61/1   Grain milling

This subclass includes:

– grain milling: production of flour, groats, meal or pellets of wheat, rye, oats, maize (corn) or other cereal grains
– manufacture of flour mixes and prepared blended flour and dough for bread, cakes, biscuits or pancakes

This subclass excludes:

– wet corn milling, see 10.62

10.61/2   Manufacture of breakfast cereals and cereals-based foods

This subclass includes:

– manufacture of cereal breakfast foods
– rice milling: production of husked, milled, polished, glazed, parboiled or converted rice; production of rice flour
– vegetable milling: production of flour or meal of dried leguminous vegetables, of roots or tubers, or of edible nuts

This subclass excludes:

– manufacture of potato flour and meal, see 10.31

10.62   Manufacture of starches and starch products

This class includes:

– manufacture of starches from rice, potatoes, maize etc.
– wet corn milling
– manufacture of glucose, glucose syrup, maltose, inulin etc.
– manufacture of gluten
– manufacture of tapioca and tapioca substitutes prepared from starch
– manufacture of corn oil

This class excludes:

– manufacture of lactose (milk sugar), see 10.51
– production of cane or beet sugar, see 10.81

10.7   Manufacture of bakery and farinaceous products

This group includes the production of bakery products, macaroni noodles and similar products.

10.71   Manufacture of bread; manufacture of fresh pastry goods and cakes

This class includes:

– manufacture of bakery products:
    ■ bread and rolls
    ■ pastry, cakes, pies, tarts, pancakes, waffles, rolls etc.

This class excludes:

– manufacture of dry bakery products, see 10.72

– *manufacture of pastas, see 10.73*

– *heating up of bakery items for immediate consumption, see division 56*

10.72   Manufacture of rusks and biscuits; manufacture of preserved pastry goods and cakes

This class includes:

– manufacture of rusks, biscuits and other dry bakery products

– manufacture of preserved pastry goods and cakes

– manufacture of snack products (cookies, crackers, pretzels etc.), whether sweet or salted

This class excludes:

– *manufacture of potato snacks, see 10.31*

10.73   Manufacture of macaroni, noodles, couscous and similar farinaceous products

This class includes:

– manufacture of pastas such as macaroni and noodles, whether or not cooked or stuffed

– manufacture of couscous

– manufacture of canned or frozen pasta products

This class excludes:

– *manufacture of prepared couscous dishes, see 10.85*

– *manufacture of soup containing pasta, see 10.89*

10.8   Manufacture of other food products

This group includes the production sugar and confectionery, prepared meals and dishes, coffee, tea and spices, as well as perishable and specialty food products.

10.81   Manufacture of sugar

This class includes:

– manufacture or refining of sugar (sucrose) and sugar substitutes from the juice of cane, beet, maple and palm

– manufacture of sugar syrups

– manufacture of molasses

– production of maple syrup and sugar

This class excludes:

– *manufacture of glucose, glucose syrup, maltose, see 10.62*

10.82   Manufacture of cocoa, chocolate and sugar confectionery

10.82/1   Manufacture of cocoa, and chocolate confectionery

This subclass includes:

– manufacture of cocoa, cocoa butter, cocoa fat, cocoa oil

– manufacture of chocolate and chocolate confectionery

This subclass excludes:

– *manufacture of white chocolate, see 10.82/2*

10.82/2   Manufacture of sugar confectionery

This subclass includes:

– manufacture of sugar confectionery: caramels, cachous, nougats, fondant, white chocolate

– manufacture of chewing gum

– preserving in sugar of fruit, nuts, fruit peels and other parts of plants

– manufacture of confectionery lozenges and pastilles

This subclass excludes:

– *manufacture of sucrose sugar, see 10.81*

10.83   Processing of tea and coffee

10.83/1   Tea processing

This subclass includes:

– blending of tea and maté

– manufacture of extracts and preparations based on tea or maté
– packing of tea including packing in tea-bags

This subclass also includes:

– manufacture of herb infusions (mint, vervain, chamomile etc.)

This subclass excludes:

– *manufacture of inulin, see 10.62*
– *manufacture of spirits, beer, wine and soft drinks, see division 11*
– *preparation of botanical products for pharmaceutical use, see 21.20*

10.83/2   **Production of coffee and coffee substitutes**

This subclass includes:

– decaffeinating and roasting of coffee
– production of coffee products:
  - ground coffee
  - soluble coffee
  - extracts and concentrates of coffee
– manufacture of coffee substitutes

This subclass excludes:

– *manufacture of inulin, see 10.62*
– *manufacture of spirits, beer, wine and soft drinks, see division 11*
– *preparation of botanical products for pharmaceutical use, see 21.20*

10.84   **Manufacture of condiments and seasonings**

This class includes:

– manufacture of spices, sauces and condiments:
  - mayonnaise
  - mustard flour and meal
  - prepared mustard etc.
– manufacture of vinegar
This class also includes:

– processing of salt into food-grade salt, e.g. iodised salt

This class excludes:

– *growing of spice crops, see 01.28*

10.85   **Manufacture of prepared meals and dishes**

This class includes the manufacture of ready-made (i.e. prepared, seasoned and cooked) meals and dishes. These dishes are processed to preserve them, such as in frozen or canned form, and are usually packaged and labelled for re-sale, i.e. this class does not include the preparation of meals for immediate consumption, such as in restaurants. To be considered a dish, these foods have to contain at least two distinct ingredients (except seasonings etc.).

This class includes:

– manufacture of meat or poultry dishes
– manufacture of fish dishes, including fish and chips
– manufacture of vegetable dishes
– manufacture of frozen or otherwise preserved pizza

This class also includes:

– manufacture of local and national dishes

This class excludes:

– *manufacture of fresh foods or foods with less than two ingredients, see corresponding class in division 10*
– *manufacture of perishable prepared foods, see 10.89*
– *retail sale of prepared meals and dishes in stores, see 47.11, 47.29*
– *wholesale of prepared meals and dishes, see 46.38*
– *activities of food service contractors, see 56.29*

10.86   **Manufacture of homogenised food preparations and dietetic food**

This class includes:

– manufacture of foods for particular nutritional uses:
  ■ infant formulae
  ■ follow-up milk and other follow-up foods
  ■ baby foods
  ■ low-energy and energy-reduced foods intended for weight control
  ■ dietary foods for special medical purposes
  ■ low-sodium foods, including low-sodium or sodium-free dietary salts
  ■ gluten-free foods
  ■ foods intended to meet the expenditure of intense muscular effort, especially for sportsmen
  ■ foods for persons suffering from carbohydrate metabolism disorders (diabetes)

10.89   **Manufacture of other food products n.e.c.**

This class includes:

– manufacture of soups and broths
– manufacture of artificial honey and caramel
– manufacture of perishable prepared foods, such as:
  ■ sandwiches
  ■ fresh (uncooked) pizza
– manufacture of food supplements and other food products n.e.c.

This class also includes:

– manufacture of yeast
– manufacture of extracts and juices of meat, fish, crustaceans or molluscs
– manufacture of non-dairy milk and cheese substitutes
– manufacture of egg products, egg albumin
– manufacture of artificial concentrates

This class excludes:

– *manufacture of perishable prepared foods of fruit and vegetables, see 10.39*
– *manufacture of frozen pizza, see 10.85*
– *manufacture of spirits, beer, wine and soft drinks, see division 11*

10.9   **Manufacture of prepared animal feeds**

10.91   **Manufacture of prepared feeds for farm animals**

This class includes:

– manufacture of prepared feeds for farm animals, including concentrated animal feed and feed supplements
– preparation of unmixed (single) feeds for farm animals

This class also includes:

– treatment of slaughter waste to produce animal feeds

This class excludes:

– *production of fishmeal for animal feed, see 10.20*
– *production of oilseed cake, see 10.41*
– *activities resulting in by-products usable as animal feed without special treatment, e.g. oilseeds (see 10.41), grain milling residues (see 10.61) etc.*

10.92   **Manufacture of prepared pet foods**

This class includes:

– manufacture of prepared feeds for pets, including dogs, cats, birds, fish etc.

This class also includes:

– treatment of slaughter waste to produce animal feeds

This class excludes:

– *production of fishmeal for animal feed, see 10.20*

– *production of oilseed cake, see 10.41*
– *activities resulting in by-products usable as animal feed without special treatment, e.g. oilseeds, (see10.41), grain milling residues (see 10.61) etc.*

11   Manufacture of beverages

This division includes the manufacture of beverages, such as non-alcoholic beverages and mineral water, manufacture of alcoholic beverages mainly through fermentation, beer and wine, and the manufacture of distilled alcoholic beverages.

This division excludes:

– *production of fruit and vegetable juices, see 10.32*
– *manufacture of milk-based drinks, see 10.51*
– *manufacture of coffee, tea and maté products, see 10.83*

11.0   Manufacture of beverages

11.01   Distilling, rectifying and blending of spirits

This class includes:

– manufacture of distilled, potable, alcoholic beverages: whisky, brandy, gin, liqueurs etc.
– manufacture of drinks mixed with distilled alcoholic beverages
– blending of distilled spirits
– production of neutral spirits

This class excludes:

– *manufacture of non-distilled alcoholic beverages, see 11.02-11.06*
– *manufacture of synthetic ethyl alcohol, see 20.14*
– *manufacture of ethyl alcohol from fermented materials, see 20.14*
– *merely bottling and labelling, see 46.34 (if performed as part of wholesale) and 82.92 (if performed on a fee or contract basis)*

11.02   Manufacture of wine from grape

This class includes:

– manufacture of wine
– manufacture of sparkling wine
– manufacture of wine from concentrated grape must

This class also includes:

– blending, purification and bottling of wine
– manufacture of low or non-alcoholic wine

This class excludes:

– *merely bottling and labelling, see 46.34 (if performed as part of wholesale) and 82.92 (if performed on a fee or contract basis)*

11.03   Manufacture of cider and other fruit wines

This class includes:

– manufacture of fermented but not distilled alcoholic beverages: sake, cider, perry and other fruit wines

This class also includes:

– manufacture of mead and mixed beverages containing fruit wines

This class excludes:

– *merely bottling and labelling, see 46.34 (if performed as part of wholesale) and 82.92 (if performed on a fee or contract basis)*

11.04   Manufacture of other non-distilled fermented beverages

This class includes:

– manufacture of vermouth and the like

This class excludes:

– *merely bottling and labelling, see 46.34 (if performed as part of wholesale) and 82.92 (if performed on a fee or contract basis)*

11.05   Manufacture of beer

This class includes:

– manufacture of malt liquors, such as beer, ale, porter and stout

This class also includes:

– manufacture of low alcohol or non-alcoholic beer

11.06   Manufacture of malt

11.07   Manufacture of soft drinks; production of mineral waters and other bottled waters

This class includes manufacture of non-alcoholic beverages (except non-alcoholic beer and wine):

– production of natural mineral waters and other bottled waters
– manufacture of soft drinks:
  ■ non-alcoholic flavoured and/or sweetened waters: lemonade, orangeade, cola, fruit drinks, tonic waters etc.

This class excludes:

– *production of fruit and vegetable juice, see 10.32*
– *manufacture of milk-based drinks, see 10.51*
– *manufacture of coffee, tea and maté products, see 10.83*
– *manufacture of alcohol-based drinks, see 11.01, 11.02, 11.03, 11.04, 11.05*
– *manufacture of non-alcoholic wine, see 11.02*
– *manufacture of non-alcoholic beer, see 11.05*
– *manufacture of ice, see 35.30*
– *merely bottling and labelling, see 46.34 (if performed as part of wholesale) and 82.92 (if performed on a fee or contract basis)*

12   **Manufacture of tobacco products**

This division includes the processing of an agricultural product, tobacco, into a form suitable for final consumption.

12.0   **Manufacture of tobacco products**

12.00   Manufacture of tobacco products

This class includes:

– manufacture of tobacco products and products of tobacco substitutes: cigarettes, fine cut tobacco, cigars, pipe tobacco, chewing tobacco, snuff
– manufacture of "homogenised" or "reconstituted" tobacco

This class also includes:

– stemming and redrying of tobacco

This class excludes:

– *growing or preliminary processing of tobacco, see 01.15, 01.63*

13   **Manufacture of textiles**

This division includes preparation and spinning of textile fibres as well as textile weaving, finishing of textiles and wearing apparel, manufacture of made-up textile articles, except apparel (e.g. household linen, blankets, rugs, cordage etc.). Growing of natural fibres is covered under division 01, while manufacture of synthetic fibres is a chemical process classified in class 20.60. Manufacture of wearing apparel is covered in division 14.

13.1   **Preparation and spinning of textile fibres**

13.10   Preparation and spinning of textile fibres

This class includes preparatory operations on textile fibres and the spinning of textile fibres. This can be done from varying raw materials, such as silk, wool, other animal, vegetable or man-made fibres, paper or glass etc.

This class includes:

– preparatory operations on textile fibres:
  ■ reeling and washing of silk
  ■ degreasing and carbonising of wool and dyeing of wool fleece
  ■ carding and combing of all kinds of animal, vegetable and man-made fibres
– spinning and manufacture of yarn or thread for weaving or sewing, for the trade or for further processing
  ■ scutching of flax
  ■ texturising, twisting, folding, cabling and dipping of synthetic or artificial filament yarns

This class also includes:

– manufacture of paper yarn

This class excludes:

– *preparatory operations carried out in combination with agriculture, see division 01*
– *retting of plants bearing vegetable textile fibres (jute, flax, coir etc.), see 01.16*
– *cotton ginning, see 01.63*
– *manufacture of synthetic or artificial fibres and tows, manufacture of single yarns (including high-tenacity yarn and yarn for carpets) of synthetic or artificial fibres, see 20.60*
– *manufacture of glass fibres, see 23.14*

13.2　　　Weaving of textiles

13.20　　　Weaving of textiles

This class includes weaving of textiles. This can be done from varying raw materials, such as silk, wool, other animal, vegetable or man-made fibres, paper or glass etc.

This class includes:

– manufacture of broad woven cotton-type, woollen-type, worsted-type or silk-type fabrics, including from mixtures of artificial or synthetic yarns (polypropylene etc.)
– manufacture of other broad woven fabrics, using flax, ramie, hemp, jute, bast fibres and special yarns

This class also includes:

– manufacture of woven pile or chenille fabrics, terry towelling, gauze etc.
– manufacture of woven fabrics of glass fibres
– manufacture of woven carbon and aramid threads
– manufacture of imitation fur by weaving

This class excludes:

– *manufacture of knitted and crocheted fabrics, see 13.91*
– *manufacture of textile floor coverings, see 13.93*
– *manufacture of non-woven fabrics, see 13.95*
– *manufacture of narrow fabrics, see 13.96*
– *manufacture of felts, see 13.99*

13.3　　　Finishing of textiles

13.30　　　Finishing of textiles

This class includes finishing of textiles and wearing apparel, i.e. bleaching, dyeing, dressing and similar activities.

This class includes:

– bleaching and dyeing of textile fibres, yarns, fabrics and textile articles, including wearing apparel
– dressing, drying, steaming, shrinking, mending, sanforising, mercerising of textiles and textile articles, including wearing apparel

This class also includes:

– bleaching of jeans
– pleating and similar work on textiles
– waterproofing, coating, rubberising, or impregnating purchased garments
– silk screen-printing on textiles and wearing apparel

This class excludes:

– *manufacture of textile fabric impregnated, coated, covered or laminated with rubber, where rubber is the chief constituent, see 22.19*

13.9　　　Manufacture of other textiles

This group includes the manufacture of products produced from textiles, except wearing apparel, such as made-up textile articles, carpets and rugs, rope, narrow woven fabrics, trimmings etc.

13.91　　　Manufacture of knitted and crocheted fabrics

This class includes:

– manufacture and processing of knitted or crocheted fabrics:
  ■ pile and terry fabrics

■ net and window furnishing type fabrics knitted on Raschel or similar machines
■ other knitted or crocheted fabrics

This class also includes:

– manufacture of imitation fur by knitting

This class excludes:

– *manufacture of net and window furnishing type fabrics of lace knitted on Raschel or similar machines, see 13.99*
– *manufacture of knitted and crocheted apparel, see 14.39*

13.92   **Manufacture of made-up textile articles, except apparel**

13.92/1   **Manufacture of soft furnishings**

This subclass includes:

– manufacture of made-up soft furnishing articles of any textile material including of knitted or crocheted fabrics:
■ cushions, pouffes, pillows, curtains, furniture covers, blinds such as Festoon, Austrian, Roman, etc.

This subclass excludes:

– *canvas blinds and sunblinds, see 13.92/2*

13.92/2   **Manufacture of canvas goods, sacks etc.**

This subclass includes:

– manufacture of made-up canvas goods, sacks, etc.:
■ tarpaulins, tents, camping goods, sails, sacks, blinds, sun blinds, machine covers, loose
■ covers for cars
■ flags, banners, pennants, etc.
■ life jackets, parachutes, etc.

13.92/3   **Manufacture of household textiles (other than soft furnishings of 13.92/1)**

This subclass includes:

– manufacture of made-up household textile articles of any textile material including of knitted and crocheted fabrics:
■ blankets including travelling rugs
■ bed, table, toilet or kitchen linen
■ quilts, eiderdowns, sleeping bags, etc.
■ valances, bedspreads, etc.
■ dust cloths, dishcloths and similar articles

This subclass also includes:

– manufacture of the textile part of electric blankets
– manufacture of hand-woven tapestries

This subclass excludes:

– *manufacture of textile articles for technical use see 13.96*

13.93   **Manufacture of carpets and rugs**

13.93/1   **Manufacture of woven or tufted carpets and rugs**

This subclass includes:

– manufacture of pile carpets, rugs and mats from wool, cotton and man-made fibres by weaving processes
– manufacture of pile carpets, rugs, mats and tiles from wool, cotton and man-made fibres by tufting processes

13.93/9   **Manufacture of carpets and rugs (other than woven or tufted) n.e.c.**

This subclass includes:

– manufacture of needle-loom and bonded fibre carpets, rugs and tiles and needlefelt underlay
– manufacture of jute carpets and mats
– manufacture of knotted carpets
– manufacture of mats and matting of coir, sisal and other hard fibres

This subclass excludes:

– *manufacture of mats and matting of plaiting materials see 16.29*
– *manufacture of floor coverings of cork, see 16.29*
– *manufacture of resilient floor coverings, such as vinyl, linoleum, see 22.23*

13.94   Manufacture of cordage, rope, twine and netting

This class includes:

- manufacture of twine, cordage, rope and cables of textile fibres or strip or the like, whether or not impregnated, coated, covered or sheathed with rubber or plastics
- manufacture of knotted netting of twine, cordage or rope
- manufacture of products of rope or netting: fishing nets, ships' fenders, unloading cushions, loading slings, rope or cable fitted with metal rings etc.

This class excludes:

- *manufacture of hairnets, see 14.19*
- *manufacture of wire rope, see 25.93*
- *manufacture of landing nets for sports fishing, see 32.30*

13.95   Manufacture of non-wovens and articles made from non-wovens, except apparel

This class includes all activities related to the manufacture of textiles or textile products, not specified elsewhere in divisions 13 or 14, involving a large number of processes and a great variety of goods produced.

13.96   Manufacture of other technical and industrial textiles

This class includes:

- manufacture of narrow woven fabrics, including fabrics consisting of warp without weft assembled by means of an adhesive
- manufacture of labels, badges etc.
- manufacture of ornamental trimmings: braids, tassels, pompons etc.
- manufacture of fabrics impregnated, coated, covered or laminated with plastics
- manufacture of metallised yarn or metallised gimped yarn, rubber thread and cord covered with textile material, textile yarn or strip covered, impregnated, coated or sheathed with rubber or plastics
- manufacture of tyre cord fabric of high-tenacity man-made yarn
- manufacture of other treated or coated fabrics: buckram and similar stiffened textile fabrics, fabrics coated with gum or amylaceous substances
- manufacture of diverse textile articles: textile wicks, incandescent gas mantles and tubular gas
- manufacture of mantle fabric, hosepiping, transmission or conveyor belts or belting (whether or not reinforced with metal or other material), bolting cloth, straining cloth
- manufacture of automotive trimmings
- manufacture of artists' canvas boards and tracing cloth

This class excludes:

- *manufacture of transmission or conveyor belts of textile fabric, yarn or cord impregnated, coated, covered or laminated with rubber, where rubber is the chief constituent, see 22.19*
- *manufacture of plates or sheets of cellular rubber or plastic combined with textiles for reinforcing purposes only, see 22.19, 22.21*
- *manufacture of cloth of woven metal wire, see 25.93*

13.99   Manufacture of other textiles n.e.c.

This class includes:

- manufacture of felt
- manufacture of tulles and other net fabrics, and of lace and embroidery, in the piece, in strips or in motifs
- manufacture of pressure sensitive cloth-tape
- manufacture of shoe-lace, of textiles
- manufacture of powder puffs and mitts

This class excludes:

- *manufacture of needle-loom felt floor coverings, see 13.93*
- *manufacture of textile wadding and articles of wadding: sanitary towels, tampons etc., see 17.22*

14   **Manufacture of wearing apparel**

This division includes all tailoring (ready-to-wear or made-to-measure), in all materials (e.g. leather, knitted and crocheted fabrics etc.), of all items of clothing (e.g. outerwear, underwear for men, women or children; work, city or casual clothing etc.) and accessories. There is no distinction made between clothing for adults and clothing for children, or between modern and traditional clothing. Division 14 also includes the fur industry (fur skins and wearing apparel).

14.1   **Manufacture of wearing apparel, except fur apparel**

This group includes manufacture of wearing apparel. The material used may be of any kind and may be coated, impregnated or rubberised.

C

14.11   Manufacture of leather clothes

This class includes:

– manufacture of wearing apparel made of leather or composition leather including leather industrial work accessories as welder's leather aprons

This class excludes:

– *manufacture of fur wearing apparel, see 14.20*
– *manufacture of leather sports gloves and sports headgear, see 32.30*
– *manufacture of fire-resistant and protective safety clothing, see 32.99*

14.12   Manufacture of workwear

This class excludes:

– *manufacture of footwear, see 15.20*
– *manufacture of fire-resistant and protective safety clothing, see 32.99*
– *repair of wearing apparel, see 95.29*

14.13   Manufacture of other outerwear

14.13/1   Manufacture of men's outerwear, other than leather clothes and workwear

This subclass includes:

– manufacture of other outerwear made of woven, knitted or crocheted fabric, non-wovens etc. for men and boys:
  ■ coats, suits, jackets, trousers etc.

This subclass also includes:

– custom tailoring
– manufacture of parts of the products listed

This subclass excludes:

– *manufacture of headgear of fur skins, see 14.19 and wearing apparel of fur skins, see 14.20*
– *manufacture of wearing apparel of rubber or plastics not assembled by stitching but merely sealed together, see 22.19, 22.29*
– *manufacture of fire-resistant and protective safety clothing, see 32.99*
– *repair of wearing apparel, see 95.29*

14.13/2   Manufacture of women's outerwear, other than leather clothes and workwear

This subclass includes:

– manufacture of other outerwear made of woven, knitted or crocheted fabric, non-wovens etc. for women and girls:
  ■ coats, suits, ensembles, jackets, trousers, skirts etc.

This subclass also includes:

– custom tailoring
– manufacture of parts of the products listed

This subclass excludes:

– *manufacture of headgear of fur skins, see 14.19 and wearing apparel of fur skins, see 14.20*
– *manufacture of wearing apparel of rubber or plastics not assembled by stitching but merely sealed together, see 22.19, 22.29*
– *manufacture of fire-resistant and protective safety clothing, see 32.99*
– *repair of wearing apparel, see 95.29*

14.14   Manufacture of underwear

14.14/1   Manufacture of men's underwear

This subclass includes:

– manufacture of underwear and nightwear made of woven, knitted or crocheted fabric, etc. for men, and boys:
  ■ shirts, T-shirts, underpants, briefs, pyjamas, dressing gowns, etc.

This subclass excludes:

– *repair of wearing apparel, see 95.29*

14.14/2   Manufacture of women's underwear

This subclass includes:

– manufacture of underwear and nightwear made of woven, knitted or crocheted fabric, lace etc. for women and girls:
  ■ blouses, lingerie, slips, brassieres, corsets, night-dresses, dressing gowns, etc.

This subclass excludes:

– *repair of wearing apparel, see 95.29*

14.19　　Manufacture of other wearing apparel and accessories n.e.c.

This class includes:

– manufacture of babies' garments, tracksuits, ski suits, swimwear etc.
– manufacture of hats and caps
– manufacture of other clothing accessories: gloves, belts, shawls, ties, cravats, hairnets etc.

This class also includes:

– manufacture of headgear of fur skins
– manufacture of footwear of textile material without applied soles
– manufacture of parts of the products listed

This class excludes:

– *manufacture of sports headgear, see 32.30*
– *manufacture of safety headgear (except sports headgear), see 32.99*
– *manufacture of fire-resistant and protective safety clothing, see 32.99*
– *repair of wearing apparel, see 95.29*

14.2　　**Manufacture of articles of fur**

14.20　　Manufacture of articles of fur

This class includes:

– manufacture of articles made of fur skins:
  - fur wearing apparel and clothing accessories
  - assemblies of fur skins such as "dropped" fur skins, plates, mats, strips etc.
  - diverse articles of fur skins: rugs, unstuffed pouffes, industrial polishing cloths

This class excludes:

– *production of raw fur skins, see 01.4, 01.70*
– *production of raw hides and skins, see 10.11*
– *manufacture of imitation furs (long-hair cloth obtained by weaving or knitting), see 13.20, 13.91*
– *manufacture of fur hats, see 14.19*
– *manufacture of apparel trimmed with fur, see 14.19*
– *dressing and dyeing of fur, see 15.11*
– *manufacture of boots or shoes containing fur parts, see 15.20*

14.3　　**Manufacture of knitted and crocheted apparel**

14.31　　Manufacture of knitted and crocheted hosiery

This class includes:

– manufacture of hosiery, including socks, tights and pantyhose

14.39　　Manufacture of other knitted and crocheted apparel

This class includes:

– manufacture of knitted or crocheted wearing apparel and other made-up articles directly into shape: pullovers, cardigans, jerseys, waistcoats and similar articles

This class excludes:

– *manufacture of knitted and crocheted fabrics, see 13.91*
– *manufacture of hosiery, see 14.31*

15　　**Manufacture of leather and related products**

This division includes dressing and dyeing of fur and the transformation of hides into leather by tanning or curing and fabricating the leather into products for final consumption. It also includes the manufacture of similar products from other materials (imitation leathers or leather substitutes), such as rubber footwear, textile luggage etc. The products made from leather substitutes are included here, since they are made in ways similar to those in which leather products are made (e.g. luggage) and are often produced in the same unit.

15.1　　**Tanning and dressing of leather; manufacture of luggage, handbags, saddlery and harness; dressing and dyeing of fur**

This group includes the manufacture of leather and fur and products thereof.

C

15.11   Tanning and dressing of leather; dressing and dyeing of fur

This class includes:

– tanning, dyeing and dressing of hides and skins
– manufacture of chamois dressed, parchment dressed, patent or metallised leathers
– manufacture of composition leather
– scraping, shearing, plucking, currying, tanning, bleaching, and dyeing of fur skins and hides with the hair on

This class excludes:

– *production of hides and skins as part of livestock farming, see 01.4*
– *production of hides and skins as part of slaughtering, see 10.11*
– *manufacture of leather apparel, see 14.11*
– *manufacture of imitation leather not based on natural leather, see 22.19, 22.29*

15.12   Manufacture of luggage, handbags and the like, saddlery and harness

This class includes:

– manufacture of luggage, handbags and the like, of leather, composition leather or any other material, such as plastic sheeting, textile materials, vulcanised fibre or paperboard, where the same technology is used as for leather
– manufacture of saddlery and harness
– manufacture of non-metallic watch bands (e.g. fabric, leather, plastic)
– manufacture of diverse articles of leather or composition leather: driving belts, packings etc.
– manufacture of shoe-lace, of leather
– manufacture of horse whips and riding crops

This class excludes:

– *manufacture of leather wearing apparel, see 14.11*
– *manufacture of leather gloves and hats, see 14.19*
– *manufacture of footwear, see 15.20*
– *manufacture of saddles for bicycles, see 30.92*
– *manufacture of precious metal watch bands, see 32.12*
– *manufacture of non-precious metal watch bands, see 32.13*
– *manufacture of linemen's safety belts and other belts for occupational use, see 32.99*

15.2   **Manufacture of footwear**

15.20   Manufacture of footwear

This class includes:

– manufacture of footwear for all purposes, of any material, by any process, including moulding (see below for exceptions)
– manufacture of leather parts of footwear: manufacture of uppers and parts of uppers, outer and inner soles, heels etc.
– manufacture of gaiters, leggings and similar articles

This class excludes:

– *manufacture of footwear of textile material without applied soles, see 14.19*
– *manufacture of wooden shoe parts (e.g. heels and lasts), see 16.29*
– *manufacture of rubber boot and shoe heels and soles and other rubber footwear parts, see 22.19*
– *manufacture of plastic footwear parts, see 22.29*
– *manufacture of ski boots, see 32.30*
– *manufacture of orthopaedic shoes, see 32.50*

16   **Manufacture of wood and of products of wood and cork, except furniture; manufacture of articles of straw and plaiting materials**

This division includes the manufacture of wooden products, such as timber, plywood, veneers, wooden containers, wooden flooring, wooden trusses, and prefabricated wooden buildings. The production processes include sawing, planing, shaping, laminating, and assembling of wood products starting from logs that are cut into bolts, or timber that may then be cut further, or shaped by lathes or other shaping tools. The timber or other transformed wood shapes may also be subsequently planed or smoothed, and assembled into finished products, such as wooden containers.

With the exception of sawmilling, this division is subdivided mainly based on the specific products manufactured.

This division does not include the manufacture of furniture (31.0), or the installation of wooden fittings and the like (43.32, 43.33, 43.39).

16.1   **Sawmilling and planing of wood**

16.10   Sawmilling and planing of wood

This class includes:

- sawing, planing and machining of wood
- slicing, peeling or chipping logs
- manufacture of wooden railway sleepers
- manufacture of unassembled wooden flooring
- manufacture of wood wool, wood flour, chips, particles

This class also includes:

- drying of wood
- impregnation or chemical treatment of wood with preservatives or other materials

This class excludes:

- *logging and production of wood in the rough, see 02.20*
- *manufacture of veneer sheets thin enough for use in plywood, boards and panels, see 16.21*
- *manufacture of shingles and shakes, beadings and mouldings, see 16.23*
- *manufacture of fire logs or pressed wood, see 16.29*

16.2   **Manufacture of products of wood, cork, straw and plaiting materials**

This group includes the manufacture of products of wood, cork, straw or plaiting materials, including basic shapes as well as assembled products.

16.21   Manufacture of veneer sheets and wood-based panels

This class includes:

- manufacture of veneer sheets thin enough to be used for veneering, making plywood or other purposes:
  - smoothed, dyed, coated, impregnated, reinforced (with paper or fabric backing)
  - made in the form of motifs
- manufacture of plywood, veneer panels and similar laminated wood boards and sheets
- manufacture of oriented strand board (OSB) and other particle board
- manufacture of medium density fibreboard (MDF) and other fibreboard
- manufacture of densified wood
- manufacture of glue laminated wood, laminated veneer wood

16.22   Manufacture of assembled parquet floors

This class includes:

- manufacture of wooden parquet floor blocks, strips etc., assembled into panels

This class excludes:

- *manufacture of unassembled wooden floors, see 16.10*

16.23   Manufacture of other builders' carpentry and joinery

This class includes:

- manufacture of wooden goods intended to be used primarily in the construction industry:
  - beams, rafters, roof struts
  - glue-laminated and metal connected, prefabricated wooden roof trusses
  - doors, windows, shutters and their frames, whether or not containing metal fittings, such as hinges, locks etc.
  - stairs, railings
  - wooden beadings and mouldings, shingles and shakes
- manufacture of prefabricated buildings, or elements thereof, predominantly of wood, e.g. saunas
- manufacture of mobile homes
- manufacture of wood partitions (except free standing)

This class excludes:

- *manufacture of kitchen cabinets, bookcases, wardrobes etc., see 31.01, 31.02, 31.09*
- *manufacture of wood partitions, free standing, see 31.01, 31.02, 31.09*

C

16.24   Manufacture of wooden containers

This class includes:

– manufacture of packing cases, boxes, crates, drums and similar packings of wood
– manufacture of pallets, box pallets and other load boards of wood
– manufacture of barrels, vats, tubs and other coopers' products of wood
– manufacture of wooden cable-drums

This class excludes:

– *manufacture of luggage, see 15.12*
– *manufacture of cases of plaiting material, see 16.29*

16.29   Manufacture of other products of wood; manufacture of articles of cork, straw and plaiting materials

This class includes:

– manufacture of various wooden products:
    ▪ wooden handles and bodies for tools, brooms, brushes
    ▪ wooden boot or shoe lasts and trees, clothes hangers
    ▪ household utensils and kitchenware of wood
    ▪ wooden statuettes and ornaments, wood marquetry, inlaid wood
    ▪ wooden cases for jewellery, cutlery and similar articles
    ▪ wooden spools, cops, bobbins, sewing thread reels and similar articles of turned wood
    ▪ other articles of wood
– natural cork processing, manufacture of agglomerated cork
– manufacture of articles of natural or agglomerated cork, including floor coverings
– manufacture of plaits and products of plaiting materials: mats, matting, screens, cases etc.
– manufacture of basket-ware and wickerwork
– manufacture of fire logs and pellets for energy, made of pressed wood or substitute materials like coffee or soybean grounds
– manufacture of wooden mirror and picture frames
– manufacture of frames for artists' canvases
– manufacture of wooden shoe parts (e.g. heels and lasts)
– manufacture of handles for umbrellas, canes and similar
– manufacture of blocks for the manufacture of smoking pipes

This class excludes:

– *manufacture of mats or matting of textile materials, see 13.92*
– *manufacture of luggage, see 15.12*
– *manufacture of wooden footwear, see 15.20*
– *manufacture of matches, see 20.51*
– *manufacture of clock cases, see 26.52*
– *manufacture of wooden spools and bobbins that are part of textile machinery, see 28.94*
– *manufacture of furniture, see 31.0*
– *manufacture of wooden toys, see 32.40*
– *manufacture of brushes and brooms, see 32.91*
– *manufacture of coffins, see 32.99*
– *manufacture of cork life preservers, see 32.99*

17   Manufacture of paper and paper products

This division includes the manufacture of pulp, paper and converted paper products. The manufacture of these products is grouped together because they constitute a series of vertically connected processes. More than one activity is often carried out in a single unit.

There are essentially three activities: The manufacture of pulp involves separating the cellulose fibres from other matter in wood, or dissolving and de-inking of used paper, and mixing in small amounts of reagents to reinforce the binding of the fibres. The manufacture of paper involves releasing pulp onto a moving wire mesh so as to form a continuous sheet. Converted paper products are made from paper and other materials by various techniques.

The paper articles may be printed (e.g. wallpaper, gift wrap etc.), as long as the printing of information is not the main purpose.

The production of pulp, paper and paperboard in bulk is included in group 17.1, while the remaining classes include the production of further-processed paper and paper products.

17.1   **Manufacture of pulp, paper and paperboard**

17.11   Manufacture of pulp

This class includes:

- manufacture of bleached, semi-bleached or unbleached paper pulp by mechanical, chemical (dissolving or non-dissolving) or semi-chemical processes
- manufacture of cotton-linters pulp
- removal of ink and manufacture of pulp from waste paper

17.12   Manufacture of paper and paperboard

This class includes:

- manufacture of paper and paperboard intended for further industrial processing

This class also includes:

- further processing of paper and paperboard:
    - coating, covering and impregnation of paper and paperboard
    - manufacture of crêped or crinkled paper
    - manufacture of laminates and foils, if laminated with paper or paperboard
- manufacture of handmade paper
- manufacture of newsprint and other printing or writing paper
- manufacture of cellulose wadding and webs of cellulose fibres
- manufacture of carbon paper or stencil paper in rolls or large sheets

This class excludes:

- *manufacture of corrugated paper and paperboard, see 17.21*
- *manufacture of further-processed articles of paper, paperboard or pulp, see 17.22, 17.23, 17.24, 17.29*
- *manufacture of coated or impregnated paper, where the coating or impregnant is the main ingredient, see class in which the manufacture of the coating or impregnant is classified*
- *manufacture of abrasive paper, see 23.91*

17.2   **Manufacture of articles of paper and paperboard**

17.21   Manufacture of corrugated paper and paperboard and of containers of paper and paperboard

17.21/1   Manufacture of corrugated paper and paperboard; manufacture of sacks and bags of paper

This subclass includes:

- manufacture of corrugated paper and paperboard
- manufacture of sacks and bags of paper

17.21/9   Manufacture of paper and paperboard containers other than sacks and bags

This subclass includes:

- manufacture of containers of corrugated paper or paperboard
- manufacture of folding paperboard containers
- manufacture of containers of solid board
- manufacture of other containers of paper and paperboard
- manufacture of office box files and similar articles

This subclass excludes:

- *manufacture of envelopes, see 17.23*
- *manufacture of moulded or pressed articles of paper pulp (e.g. boxes for packing eggs, moulded pulp paper plates), see 17.29*

17.22   Manufacture of household and sanitary goods and of toilet requisites

This class includes:

- manufacture of household and personal hygiene paper and cellulose wadding products:
    - cleansing tissues
    - handkerchiefs, towels, serviettes
    - toilet paper
    - sanitary towels and tampons, napkins and napkin liners for babies
    - cups, dishes and trays
- manufacture of textile wadding and articles of wadding: sanitary towels, tampons etc.

This class excludes:

- *manufacture of cellulose wadding, see 17.12*

17.23 Manufacture of paper stationery

This class includes:

- manufacture of printing and writing paper ready for use
- manufacture of computer printout paper ready for use
- manufacture of self-copy paper ready for use
- manufacture of duplicator stencils and carbon paper ready for use
- manufacture of gummed or adhesive paper ready for use
- manufacture of envelopes and letter-cards
- manufacture of educational and commercial stationery (notebooks, binders, registers, accounting books, business forms etc.), when the printed information is not the main characteristic
- manufacture of boxes, pouches, wallets and writing compendiums containing an assortment of paper stationery

This class excludes:

- *printing on paper products, see 18.1*

17.24 Manufacture of wallpaper

This class includes:

- manufacture of wallpaper and similar wall coverings, including vinyl-coated and textile wallpaper

This class excludes:

- *manufacture of paper or paperboard in bulk, see 17.12*
- *manufacture of plastic wall paper, see 22.29*

17.29 Manufacture of other articles of paper and paperboard

This class includes:

- manufacture of labels
- manufacture of filter paper and paperboard
- manufacture of paper and paperboard bobbins, spools, cops etc.
- manufacture of egg trays and other moulded pulp packaging products etc.
- manufacture of paper novelties
- manufacture of paper or paperboard cards for use on Jacquard machines

This class excludes:

- *manufacture of playing cards, see 32.40*
- *manufacture of games and toys of paper or paperboard, see 32.40*

18 Printing and reproduction of recorded media

This division includes printing of products, such as newspapers, books, periodicals, business forms, greeting cards, and other materials, and associated support activities, such as bookbinding, plate-making services, and data imaging. The support activities included here are an integral part of the printing industry, and a product (a printing plate, a bound book, or a computer disk or file) that is an integral part of the printing industry is almost always provided by these operations.

Processes used in printing include a variety of methods for transferring an image from a plate, screen or computer file to a medium, such as paper, plastics, metal, textile articles, or wood. The most prominent of these methods entails the transfer of the image from a plate or screen to the medium through lithographic, gravure, screen or flexographic printing. Often a computer file is used to directly ''drive'' the printing mechanism to create the image or electrostatic and other types of equipment (digital or non-impact printing).

Though printing and publishing can be carried out by the same unit (a newspaper, for example), it is less and less the case that these distinct activities are carried out in the same physical location.

This division also includes the reproduction of recorded media, such as compact discs, video recordings, software on discs or tapes, records etc.

This division excludes publishing activities (see section J).

18.1 Printing and service activities related to printing

This group includes printing of products, such as newspapers, books, periodicals, business forms, greeting cards, and other materials, and associated support activities, such as bookbinding, plate-making services, and data imaging. Printing can be done using various techniques and on different materials.

C

18.11   Printing of newspapers

This class also includes:

– printing of other periodicals, appearing at least four times a week

This class excludes:

– *publishing of printed matter, see 58.1*
– *photocopying of documents, see 82.19*

18.12   Other printing

18.12/1   Manufacture of printed labels

This subclass includes:

– printing on labels or tags (lithographic, gravure printing, flexographic printing, other)

18.12/9   Printing (other than printing of newspapers and printing on labels and tags) n.e.c.

This subclass includes:

– printing of magazines and other periodicals, appearing less than four times a week
– printing of books and brochures, music and music manuscripts, maps, atlases, posters, advertising catalogues, prospectuses and other printed advertising, postage stamps, taxation stamps, documents of title, cheques and other security papers, smart cards, albums, diaries, calendars and other commercial printed matter, personal stationery and other printed matter by letterpress, offset, photogravure, flexographic, screen printing and other printing presses, duplication machines, computer printers, embossers etc., including quick printing
– printing directly onto textiles, plastic, glass, metal, wood and ceramics (except silk-screen printing on textiles and wearing apparel)

The material printed is typically copyrighted

This subclass excludes:

– *silk screen-printing on textiles and wearing apparel, see 13.30*
– *manufacture of stationery (notebooks, binders, registers, accounting books, business forms etc.), when the printed information is not the main characteristic, see 17.23*
– *publishing of printed matter, see 58.1*

18.13   Pre-press and pre-media services

This class includes:

– composing, typesetting, phototypesetting, pre-press data input including scanning and optical character recognition, electronic make-up
– preparation of data files for multi-media (printing on paper, CD-ROM, Internet) applications
– plate-making services including image setting and plate setting (for letterpress and offset printing processes)
– cylinder preparation: engraving or etching of cylinders for gravure printing
– plate processing: "computer to plate" CTP (also photopolymer plates),
– preparation of plates and dies for relief stamping or printing
– preparation of:
  ■ artistic works of a technical character, such as lithographic stones and wood blocks
  ■ presentation of media, e.g. overhead foils and other forms of presentation
  ■ sketches, layouts, dummies, etc.
  ■ production of proofs

This class excludes:

– *specialised design activities, see 74.10*

18.14   Binding and related services

This class includes:

– trade binding, sample mounting and post press services in support of printing activities, e.g. trade binding and finishing of books, brochures, magazines, catalogues, etc., by folding, cutting and trimming, assembling, stitching, thread sewing, adhesive binding, cutting and cover laying, gluing, collating, basting, gold stamping; spiral binding and plastic wire binding
– binding and finishing of printed paper or printed cardboard, by folding, stamping, drilling, punching, perforating, embossing, sticking, gluing, laminating
– finishing services for CD-ROMs
– mailing finishing services such as customisation, envelope preparation
– other finishing activities such as die, sinking or stamping, Braille copying

C

18.2	**Reproduction of recorded media**
18.20	Reproduction of recorded media
18.20/1	Reproduction of sound recording

This subclass includes:

– reproduction from master copies of gramophone records, compact discs and tapes with music or other sound recordings

This subclass excludes:

– *reproduction of printed matter, see 18.11, 18.12*
– *production of master copies for records or audio material, see 59.20*

18.20/2	Reproduction of video recording

This subclass includes:

– reproduction from master copies of compact discs and tapes with motion pictures and other video recordings

This subclass excludes:

– *reproduction of printed matter, see 18.11, 18.12*
– *production and distribution of motion pictures, video tapes and movies on DVD or similar media, see 59.11, 59.12, 59.13*
– *reproduction of motion picture film for theatrical distribution, see 59.12*

18.20/3	Reproduction of computer media

This subclass includes:

– reproduction from master copies of software and data on discs and tapes

This subclass excludes:

– *reproduction of printed matter, see 18.11, 18.12*
– *publishing of software, see 58.2*

19	**Manufacture of coke and refined petroleum products**

This division includes the transformation of crude petroleum and coal into usable products. The dominant process is petroleum refining which involves the separation of crude petroleum into component products through such techniques as cracking and distillation.

This division also includes the manufacture for own account of characteristic products (e.g. coke, butane, propane, petrol, kerosene, fuel oil etc.) as well as processing services (e.g. custom refining).

This division includes the manufacture of gases such as ethane, propane and butane as products of petroleum refineries.

Not included, is the manufacture of such gases in other units (20.14), manufacture of industrial gases (20.11), extraction of natural gas (methane, ethane, butane or propane) (06.20), and manufacture of fuel gas, other than petroleum gases (e.g. coal gas, water gas, producer gas, gasworks gas) (35.21).

The manufacture of petrochemicals from refined petroleum is classified in division 20.

19.1	**Manufacture of coke oven products**
19.10	Manufacture of coke oven products

This class includes:

– operation of coke ovens
– production of coke and semi-coke
– production of pitch and pitch coke
– production of coke oven gas
– production of crude coal and lignite tars
– agglomeration of coke

This class excludes:

– *manufacture of coal fuel briquettes, see 19.20*

19.2	**Manufacture of refined petroleum products**
19.20	Manufacture of refined petroleum products
19.20/1	Mineral oil refining

This subclass includes the manufacture of liquid or gaseous fuels or other products from crude petroleum, bituminous minerals or their fractionation products. Petroleum refining involves one or more of the following activities: fractionation; straight distillation of crude oil; and cracking.

This subclass includes:

- production of motor fuel: gasoline, kerosene etc.
- production of fuel: light, medium and heavy fuel oil, refinery gases such as ethane, propane, butane etc.
- manufacture of oil-based lubricating oils or greases, including from waste oil
- manufacture of products for the petrochemical industry and for the manufacture of road coverings
- manufacture of various products: white spirit, vaseline, paraffin wax, petroleum jelly etc.
- manufacture of petroleum briquettes
- blending of biofuels, i.e. blending of alcohols with petroleum (e.g. gasohol)

This subclass also includes:

- manufacture of peat briquettes
- manufacture of hard-coal and lignite fuel briquettes

19.20/9 Other treatment of petroleum products (excluding mineral oil refining/petrochemicals manufacture)

This subclass includes:

- manufacture of oil-based lubricating oils or greases, including from waste oil

20 Manufacture of chemicals and chemical products

This division includes the transformation of organic and inorganic raw materials by a chemical process and the formation of products. It distinguishes the production of basic chemicals that constitute the first industry group from the production of intermediate and end products produced by further processing of basic chemicals that make up the remaining industry classes.

20.1 Manufacture of basic chemicals, fertilisers and nitrogen compounds, plastics and synthetic rubber in primary forms

This group includes the manufacture of basic chemical products, fertilisers and associated nitrogen compounds, as well as plastics and synthetic rubber in primary forms.

20.11 Manufacture of industrial gases

This class includes:

- manufacture of liquefied or compressed inorganic industrial or medical gases:
  - elemental gases
  - liquid or compressed air
  - refrigerant gases
  - mixed industrial gases
  - inert gases such as carbon dioxide
  - isolating gases

This class excludes:

- *extraction of methane, ethane, butane or propane, see 06.20*
- *manufacture of fuel gases such as ethane, butane or propane in a petroleum refinery, see 19.20*
- *manufacture of gaseous fuels from coal, waste etc., see 35.21*

20.12 Manufacture of dyes and pigments

This class includes:

- manufacture of dyes and pigments from any source in basic form or as concentrate

This class also includes:

- manufacture of products of a kind used as fluorescent brightening agents or as luminophores

This class excludes:

- *manufacture of prepared dyes and pigments, see 20.30*

20.13 Manufacture of other inorganic basic chemicals

This class includes the manufacture of chemicals using basic processes. The output of these processes are usually separate chemical elements or separate chemically-defined compounds.

This class includes:

- manufacture of chemical elements (except industrial gases and basic metals)
- manufacture of inorganic acids except nitric acid
- manufacture of alkalis, lyes and other inorganic bases except ammonia
- manufacture of other inorganic compounds

– roasting of iron pyrites
– manufacture of distilled water

This class also includes:

– enrichment of uranium and thorium ores

This class excludes:

– *manufacture of industrial gases, see 20.11*
– *manufacture of nitrogenous fertilisers and nitrogen compounds, see 20.15*
– *manufacture of ammonia, see 20.15*
– *manufacture of ammonium chloride, see 20.15*
– *manufacture of nitrites and nitrates of potassium, see 20.15*
– *manufacture of ammonium carbonates, see 20.15*
– *manufacture of aromatic distilled water, see 20.53*
– *manufacture of basic metals, see division 24*

20.14　　**Manufacture of other organic basic chemicals**

This class includes the manufacture of chemicals using basic processes, such as thermal cracking and distillation. The output of these processes are usually separate chemical elements or separate chemically defined compounds.

This class includes:

– manufacture of basic organic chemicals:
  - acyclic hydrocarbons, saturated and unsaturated
  - cyclic hydrocarbons, saturated and unsaturated
  - acyclic and cyclic alcohols
  - mono- and polycarboxylic acids, including acetic acid
  - other oxygen-function compounds, including aldehydes, ketones, quinones and dual or poly oxygen-function compounds
  - synthetic glycerol
  - nitrogen-function organic compounds, including amines
  - fermentation of sugarcane, corn or similar to produce alcohol and esters
  - other organic compounds, including wood distillation products (e.g. charcoal) etc.
– manufacture of synthetic aromatic products
– distillation of coal tar

This class excludes:

– *manufacture of plastics in primary forms, see 20.16*
– *manufacture of synthetic rubber in primary forms, see 20.17*
– *manufacture of crude glycerol, see 20.41*
– *manufacture of natural essential oils, see 20.53*
– *manufacture of salicylic and O-acetylsalicylic acids, see 21.10*

20.15　　**Manufacture of fertilisers and nitrogen compounds**

This class includes:

– manufacture of fertilisers:
  - straight or complex nitrogenous, phosphatic or potassic fertilisers
  - urea, crude natural phosphates and crude natural potassium salts
– manufacture of associated nitrogen products:
  - nitric and sulphonitric acids, ammonia, ammonium chloride, ammonium carbonate, nitrites and nitrates of potassium

This class also includes:

– manufacture of potting soil with peat as main constituent
– manufacture of potting soil mixtures of natural soil, sand, clays and minerals

This class excludes:

– *mining of guano, see 08.91*
– *manufacture of agrochemical products, such as pesticides, see 20.20*

20.16　　**Manufacture of plastics in primary forms**

This class includes the manufacture of resins, plastics materials, and non-vulcanisable thermoplastic elastomers, the mixing and blending of resins on a custom basis, as well as the manufacture of non-customised synthetic resins.

This class includes:

– manufacture of plastics in primary forms:
  - polymers, including those of ethylene, propylene, styrene, vinyl chloride, vinyl acetate and acrylics
  - polyamides
  - phenolic and epoxide resins and polyurethanes
  - alkyd and polyester resins and polyethers
  - silicones
  - ion-exchangers based on polymers

This class also includes:

– manufacture of cellulose and its chemical derivatives

This class excludes:

– *manufacture of artificial and synthetic fibres, filaments and yarn, see 20.60*
– *shredding of plastic products, see 38.32*

20.17　　**Manufacture of synthetic rubber in primary forms**

This class includes:

– manufacture of synthetic rubber in primary forms:
  - synthetic rubber
  - factice
– manufacture of mixtures of synthetic rubber and natural rubber or rubber-like gums (e.g. balata)

20.2　　**Manufacture of pesticides and other agrochemical products**

20.20　　**Manufacture of pesticides and other agrochemical products**

This class includes:

– manufacture of insecticides, rodenticides, fungicides, herbicides, acaricides, molluscicides, biocides
– manufacture of anti-sprouting products, plant growth regulators
– manufacture of disinfectants (for agricultural and other use)
– manufacture of other agrochemical products n.e.c.

This class excludes:

– *manufacture of fertilisers and nitrogen compounds, see 20.15*

20.3　　**Manufacture of paints, varnishes and similar coatings, printing ink and mastics**

20.30　　**Manufacture of paints, varnishes and similar coatings, printing ink and mastics**

20.30/1　　**Manufacture of paints, varnishes and similar coatings, mastics and sealants**

This subclass includes:

– manufacture of paints and varnishes, enamels or lacquers
– manufacture of prepared pigments and dyes, opacifiers and colours
– manufacture of vitrifiable enamels and glazes and engobes and similar preparations
– manufacture of mastics
– manufacture of caulking compounds and similar non-refractory filling or surfacing preparations
– manufacture of organic composite solvents and thinners
– manufacture of prepared paint or varnish removers

This subclass excludes:

– *manufacture of dyestuffs and pigments, see 20.12*

20.30/2　　**Manufacture of printing ink**

This subclass includes:

– manufacture of printing ink

This subclass excludes:

– *manufacture of writing and drawing ink, see 20.59*

20.4　　**Manufacture of soap and detergents, cleaning and polishing preparations, perfumes and toilet preparations**

20.41　　Manufacture of soap and detergents, cleaning and polishing preparations

C

20.41/1   Manufacture of soap and detergents

This subclass includes:

- manufacture of organic surface-active agents
- manufacture of paper, wadding, felt etc. coated or covered with soap or detergent
- manufacture of crude glycerol
- manufacture of soap, except cosmetic soap
- manufacture of surface-active preparations:
    - washing powders in solid or liquid form and detergents
    - dish-washing preparations
    - textile softeners

This subclass excludes:

- *manufacture of separate, chemically defined compounds, see 20.13, 20.14*
- *manufacture of glycerol, synthesised from petroleum products, see 20.14*
- *manufacture of cosmetic soap, see 20.42*

20.41/2   Manufacture of cleaning and polishing preparations

This subclass includes:

- manufacture of cleaning and polishing products:
    - preparations for perfuming or deodorising rooms
    - artificial waxes and prepared waxes
    - polishes and creams for leather
    - polishes and creams for wood
    - polishes for coachwork, glass and metal
    - scouring pastes and powders, including paper, wadding etc. coated or covered with these

This subclass excludes:

- *manufacture of separate, chemically defined compounds, see 20.13, 20.14*
- *manufacture of glycerol, synthesised from petroleum products, see 20.14*
- *manufacture of cosmetic soap, see 20.42*

20.42   Manufacture of perfumes and toilet preparations

This class includes:

- manufacture of perfumes and toilet preparations:
    - perfumes and toilet water
    - beauty and make-up preparations
    - sunburn prevention and suntan preparations
    - manicure and pedicure preparations
    - shampoos, hair lacquers, waving and straightening preparations
    - dentifrices and preparations for oral hygiene, including denture fixative preparations
    - shaving preparations, including pre-shave and aftershave preparations
    - deodorants and bath salts
    - depilatories
- manufacture of cosmetic soap

This class excludes:

- *extraction and refining of natural essential oils, see 20.53*

20.5   Manufacture of other chemical products

This group includes the manufacture of explosives and pyrotechnic products, glues, essential oils and chemical products n.e.c., e.g. photographic chemical material (including film and sensitised paper), composite diagnostic preparations etc.

20.51   Manufacture of explosives

This class includes:

- manufacture of propellant powders
- manufacture of explosives and pyrotechnic products, including percussion caps, detonators, signalling flares etc.

This class also includes:

- manufacture of matches

C

20.52   Manufacture of glues

This class includes:

– manufacture of glues and prepared adhesives, including rubber-based glues and adhesives

This class excludes:

– *manufacture of gelatines and its derivates, see 20.59*

20.53   Manufacture of essential oils

This class includes:

– manufacture of extracts of natural aromatic products
– manufacture of resinoids
– manufacture of mixtures of odoriferous products for the manufacture of perfumes or food

This class excludes:

– *manufacture of synthetic aromatic products, see 20.14*
– *manufacture of perfumes and toilet preparations, see 20.42*

20.59   Manufacture of other chemical products n.e.c.

This class includes:

– manufacture of photographic plates, films, sensitised paper and other sensitised unexposed materials
– manufacture of chemical preparations for photographic uses
– manufacture of gelatine and its derivatives
– manufacture of various chemical products:
  - peptones, peptone derivatives, other protein substances and their derivatives n.e.c.
  - chemically modified oils and fats
  - materials used in the finishing of textiles and leather
  - powders and pastes used in soldering, brazing or welding
  - substances used to pickle metal
  - prepared additives for cements
  - activated carbon, lubricating oil additives, prepared rubber accelerators, catalysts and other chemical products for industrial use
  - anti-knock preparations, antifreeze preparations
  - liquids for hydraulic transmission
  - composite diagnostic or laboratory reagents

This class also includes:

– manufacture of writing and drawing ink

This class excludes:

– *manufacture of chemically defined products in bulk, see 20.13, 20.14*
– *manufacture of distilled water, 20.13*
– *manufacture of other organic basic chemicals, see 20.14*
– *manufacture of printing ink, see 20.30*
– *manufacture of asphalt-based adhesives, see 23.99*

20.6   Manufacture of man-made fibres

20.60   Manufacture of man-made fibres

This class includes:

– manufacture of synthetic or artificial filament tow
– manufacture of synthetic or artificial staple fibres, not carded, combed or otherwise processed for spinning
– manufacture of synthetic or artificial filament yarn, including high-tenacity yarn
– manufacture of synthetic or artificial monofilament or strip

This class excludes:

– *spinning of synthetic or artificial fibres, see 13.10*
– *manufacture of yarns made of man-made staple, see 13.10*

21   Manufacture of basic pharmaceutical products and pharmaceutical preparations

This division includes the manufacture of basic pharmaceutical products and pharmaceutical preparations. This includes also the manufacture of medicinal chemical and botanical products.

21.1   **Manufacture of basic pharmaceutical products**

21.10   Manufacture of basic pharmaceutical products

This class includes:

- manufacture of medicinal active substances to be used for their pharmacological properties in the manufacture of medicaments: antibiotics, basic vitamins, salicylic and O-acetylsalicylic acids etc.
- processing of blood

This class also includes:

- manufacture of chemically pure sugars
- processing of glands and manufacture of extracts of glands etc.

21.2   **Manufacture of pharmaceutical preparations**

21.20   Manufacture of pharmaceutical preparations

This class includes:

- manufacture of medicaments:
  - antisera and other blood fractions
  - vaccines
  - diverse medicaments, including homeopathic preparations
- manufacture of chemical contraceptive products for external use and hormonal contraceptive medicaments
- manufacture of medical diagnostic preparations, including pregnancy tests
- manufacture of radioactive in-vivo diagnostic substances
- manufacture of biotech pharmaceuticals

This class also includes:

- manufacture of medical impregnated wadding, gauze, bandages, dressings, etc.
- preparation of botanical products (grinding, grading, milling) for pharmaceutical use

This class excludes:

- *manufacture of herb infusions (mint, vervain, chamomile etc.), see 10.83*
- *manufacture of dental fillings and dental cement, see 32.50*
- *manufacture of bone reconstruction cements, see 32.50*
- *manufacture of surgical drapes, see 32.50*
- *wholesale of pharmaceuticals, see 46.46*
- *retail sale of pharmaceuticals, see 47.73*
- *research and development for pharmaceuticals and biotech pharmaceuticals, see 72.1*
- *packaging of pharmaceuticals, see 82.92*

22   **Manufacture of rubber and plastic products**

This division includes the manufacture of rubber and plastics products.

This division is characterised by the raw materials used in the manufacturing process. However, this does not imply that the manufacture of all products made of these materials is classified here.

22.1   **Manufacture of rubber products**

This group includes the manufacture of rubber products.

22.11   Manufacture of rubber tyres and tubes; retreading and rebuilding of rubber tyres

This class includes:

- manufacture of rubber tyres for vehicles, equipment, mobile machinery, aircraft, toy, furniture and other uses:
  - pneumatic tyres
  - solid or cushion tyres
- manufacture of inner tubes for tyres
- manufacture of interchangeable tyre treads, tyre flaps, "camelback" strips for retreading tyres etc.
- tyre rebuilding and retreading

This class excludes:

- *manufacture of tube repair materials, see 22.19*
- *tyre and tube repair, fitting or replacement, see 45.20*

C

| 22.19 | Manufacture of other rubber products |

This class includes:

– manufacture of other products of natural or synthetic rubber, unvulcanised, vulcanised or hardened:
  ▪ rubber plates, sheets, strip, rods, profile shapes
  ▪ tubes, pipes and hoses
  ▪ rubber conveyor or transmission belts or belting
  ▪ rubber hygienic articles: sheath contraceptives, teats, hot water bottles etc.
  ▪ rubber articles of apparel (if only sealed together, not sewn)
  ▪ rubber sole and other rubber parts of footwear
  ▪ rubber thread and rope
  ▪ rubberised yarn and fabrics
  ▪ rubber rings, fittings and seals
  ▪ rubber roller coverings
  ▪ inflatable rubber mattresses
  ▪ inflatable balloons
– manufacture of rubber brushes
– manufacture of hard rubber pipe stems
– manufacture of hard rubber combs, hair pins, hair rollers, and similar

This class also includes:

– manufacture of rubber repair materials
– manufacture of textile fabric impregnated, coated, covered or laminated with rubber, where rubber is the chief constituent
– manufacture of rubber waterbed mattresses
– manufacture of rubber bathing caps and aprons
– manufacture of rubber wet suits and diving suits
– manufacture of rubber sex articles

This class excludes:

– *manufacture of tyre cord fabrics, see 13.96*
– *manufacture of apparel of elastic fabrics, see 14.14, 14.19*
– *manufacture of rubber footwear, see 15.20*
– *manufacture of glues and adhesives based on rubber, see 20.52*
– *manufacture of "camelback" strips, see 22.11*
– *manufacture of inflatable rafts and boats, see 30.11, 30.12*
– *manufacture of mattresses of uncovered cellular rubber, see 31.03*
– *manufacture of rubber sports requisites, except apparel, see 32.30*
– *manufacture of rubber games and toys (including children's wading pools, inflatable children rubber boats, inflatable rubber animals, balls and the like), see 32.40*
– *reclaiming of rubber, see 38.32*

| 22.2 | **Manufacture of plastics products** |

This group comprises processing new or spent (i.e., recycled) plastics resins into intermediate or final products, using such processes as compression moulding; extrusion moulding; injection moulding; blow moulding; and casting. For most of these, the production process is such that a wide variety of products can be made.

| 22.21 | Manufacture of plastic plates, sheets, tubes and profiles |

This class includes:

– manufacture of semi-manufactures of plastic products:
  ▪ plastic plates, sheets, blocks, film, foil, strip etc. (whether self-adhesive or not)
– manufacture of finished plastic products:
  ▪ plastic tubes, pipes and hoses; hose and pipe fittings
– cellophane film or sheet

This class excludes:

– *manufacture of plastics in primary forms, see 20.16*
– *manufacture of articles of synthetic or natural rubber, see 22.1*

C

22.22   Manufacture of plastic packing goods

This class includes:

– manufacture of plastic articles for the packing of goods:
  - plastic bags, sacks, containers, boxes, cases, carboys, bottles etc.

This class excludes:

– *manufacture of luggage and handbags of plastic, see 15.12*

22.23   Manufacture of builders' ware of plastic

This class includes:

– manufacture of builders' plastics ware:
  - plastic doors, windows, frames, shutters, blinds, skirting boards
  - tanks, reservoirs
  - plastic floor, wall or ceiling coverings in rolls or in the form of tiles etc.
  - plastic sanitary ware like plastic baths, shower-baths, washbasins, lavatory pans, flushing cisterns etc.
– manufacture of resilient floor coverings, such as vinyl, linoleum etc.
– manufacture of artificial stone (e.g. cultured marble)

22.29   Manufacture of other plastic products

This class includes:

– manufacture of plastic tableware, kitchenware and toilet articles
– manufacture of diverse plastic products:

  - plastic headgear, insulating fittings, parts of lighting fittings, office or school supplies, articles of apparel (if only sealed together, not sewn), fittings for furniture, statuettes, transmission and conveyer belts, self-adhesive tapes of plastic, plastic shoe lasts, plastic cigar and cigarette holders, combs, plastics hair curlers, plastics novelties, etc.

This class excludes:

– *manufacture of plastic luggage, see 15.12*
– *manufacture of plastic footwear, see 15.20*
– *manufacture of plastic furniture, see 31.01, 31.02, 31.09*
– *manufacture of mattresses of uncovered cellular plastic, see 31.03*
– *manufacture of plastic sports requisites, see 32.30*
– *manufacture of plastic games and toys, see 32.40*
– *manufacture of plastic medical and dental appliances, see 32.50*
– *manufacture of plastic ophthalmic goods, see 32.50*
– *manufacture of plastics hard hats and other personal safety equipment of plastics, see 32.99*

23   **Manufacture of other non-metallic mineral products**

This division includes manufacturing activities related to a single substance of mineral origin. This division includes the manufacture of glass and glass products (e.g. flat glass, hollow glass, fibres, technical glassware etc.), ceramic products, tiles and baked clay products, and cement and plaster, from raw materials to finished articles. The manufacture of shaped and finished stone and other mineral products is also included in this division.

23.1   **Manufacture of glass and glass products**

This group includes glass in all its forms, made by any process, and articles of glass.

23.11   Manufacture of flat glass

This class includes:

– manufacture of flat glass, including wired, coloured or tinted flat glass

23.12   Shaping and processing of flat glass

This class includes:

– manufacture of toughened or laminated flat glass
– manufacture of glass mirrors
– manufacture of multiple-walled insulating units of glass

23.13   Manufacture of hollow glass

This class includes:

– manufacture of bottles and other containers of glass or crystal

– manufacture of drinking glasses and other domestic glass or crystal articles

This class excludes:

– *manufacture of glass toys, see 32.40*

23.14   Manufacture of glass fibres

This class includes:

– manufacture of glass fibres, including glass wool and non-woven products thereof

This class excludes:

– *manufacture of woven fabrics of glass yarn, see 13.20*
– *manufacture of fibre optic cable for data transmission or live transmission of images, see 27.31*

23.19   Manufacture and processing of other glass, including technical glassware

This class includes:

– manufacture of laboratory, hygienic or pharmaceutical glassware
– manufacture of clock or watch glasses, optical glass and optical elements not optically worked
– manufacture of glassware used in imitation jewellery
– manufacture of glass insulators and glass insulating fittings
– manufacture of glass envelopes for lamps
– manufacture of glass figurines
– manufacture of glass paving blocks
– manufacture of glass in rods or tubes

This class excludes:

– *manufacture of optical elements optically worked, see 26.70*
– *manufacture of syringes and other medical laboratory equipment, see 32.50*

23.2   Manufacture of refractory products

23.20   Manufacture of refractory products

This class includes the manufacture of intermediate products from mined or quarried non-metallic minerals, such as sand, gravel, stone or clay.

This class includes:

– manufacture of refractory mortars, concretes etc.
– manufacture of refractory ceramic goods:
    - heat-insulating ceramic goods of siliceous fossil meals
    - refractory bricks, blocks and tiles etc.
    - retorts, crucibles, muffles, nozzles, tubes, pipes etc.

This class also includes:

– manufacture of refractory articles containing magnesite, dolomite or chromite

23.3   Manufacture of clay building materials

23.31   Manufacture of ceramic tiles and flags

This class includes:

– manufacture of non-refractory ceramic hearth or wall tiles, mosaic cubes etc.
– manufacture of non-refractory ceramic flags and paving

This class excludes:

– *manufacture of artificial stone (e.g. cultured marble), see 22.23*
– *manufacture of refractory ceramic products, see 23.20*
– *manufacture of ceramic bricks and roofing tiles, see 23.32*

23.32   Manufacture of bricks, tiles and construction products, in baked clay

This class includes:

– manufacture of structural non-refractory clay building materials:
    - manufacture of ceramic bricks, roofing tiles, chimney pots, pipes, conduits etc.
– manufacture of flooring blocks in baked clay

This class excludes:

- *manufacture of refractory ceramic products, see 23.20*
- *manufacture of non-structural non-refractory ceramic products, see 23.4*

23.4   Manufacture of other porcelain and ceramic products

This group includes the manufacture of final products from mined or quarried non-metallic minerals, such as sand, gravel, stone or clay.

23.41   Manufacture of ceramic household and ornamental articles

This class includes:

- manufacture of ceramic tableware and other domestic or toilet articles
- manufacture of statuettes and other ornamental ceramic articles

This class excludes:

- *manufacture of imitation jewellery, see 32.13*
- *manufacture of ceramic toys, see 32.40*

23.42   Manufacture of ceramic sanitary fixtures

This class includes:

- manufacture of ceramic sanitary fixtures, e.g. sinks, baths, bidets, water closet pans etc.
- manufacture of other ceramic fixtures

This class excludes:

- *manufacture of refractory ceramic goods, see 23.20*
- *manufacture of ceramic building materials, see 23.3*

23.43   Manufacture of ceramic insulators and insulating fittings

This class includes:

- manufacture of electrical insulators and insulating fittings of ceramics

This class excludes:

- *manufacture of refractory ceramic goods, see 23.20*

23.44   Manufacture of other technical ceramic products

This class includes:

- manufacture of ceramic and ferrite magnets
- manufacture of ceramic laboratory, chemical and industrial products

This class excludes:

- *manufacture of artificial stone (e.g. cultured marble), see 22.23*
- *manufacture of refractory ceramic goods, see 23.20*
- *manufacture of ceramic building materials, see 23.3*

23.49   Manufacture of other ceramic products

This class includes:

- manufacture of ceramic pots, jars and similar articles of a kind used for conveyance or packing of goods
- manufacture of ceramic products n.e.c.

This class excludes:

- *manufacture of ceramic sanitary fixtures, see 23.42*
- *manufacture of artificial teeth, see 32.50*

23.5   Manufacture of cement, lime and plaster

23.51   Manufacture of cement

This class includes:

- manufacture of clinkers and hydraulic cements, including Portland, aluminous cement, slag cement and superphosphate cements

This class excludes:

- *manufacture of refractory mortars, concrete etc., see 23.20*

– manufacture of ready-mixed and dry-mix concrete and mortars, see 23.63, 23.64
– manufacture of articles of cement, see 23.69
– manufacture of cements used in dentistry, see 32.50

23.52 Manufacture of lime and plaster

This class includes:

– manufacture of quicklime, slaked lime and hydraulic lime
– manufacture of plasters of calcined gypsum or calcined sulphate

This class also includes:

– manufacture of calcined dolomite

This class excludes:

– manufacture of articles of plaster, see 23.62, 23.69

23.6 Manufacture of articles of concrete, cement and plaster

23.61 Manufacture of concrete products for construction purposes

This class includes:

– manufacture of precast concrete, cement or artificial stone articles for use in construction:
  ▪ tiles, flagstones, bricks, boards, sheets, panels, pipes, posts etc.
– manufacture of prefabricated structural components for building or civil engineering of cement, concrete or artificial stone

23.62 Manufacture of plaster products for construction purposes

This class includes:

– manufacture of plaster articles for use in construction:
  ▪ boards, sheets, panels etc.

23.63 Manufacture of ready-mixed concrete

This class includes:

– manufacture of ready-mix and dry-mix concrete and mortars

This class excludes:

– manufacture of refractory cements, see 23.20

23.64 Manufacture of mortars

This class includes:

– manufacture of powdered mortars

This class excludes:

– manufacture of refractory mortars, see 23.20
– manufacture of dry-mixed concrete and mortars, see 23.63

23.65 Manufacture of fibre cement

This class includes:

– manufacture of building materials of vegetable substances (wood wool, straw, reeds, rushes) agglomerated with cement, plaster or other mineral binder
– manufacture of articles of asbestos-cement or cellulose fibre-cement or the like:
  ▪ corrugated sheets, other sheets, panels, tiles, tubes, pipes, reservoirs, troughs, basins, sinks, jars, furniture, window frames etc.

23.69 Manufacture of other articles of concrete, plaster and cement

This class includes:

– manufacture of other articles of concrete, plaster, cement or artificial stone:
  ▪ statuary, furniture, bas- and haut-reliefs, vases, flowerpots etc.

23.7 Cutting, shaping and finishing of stone

23.70 Cutting, shaping and finishing of stone

This class includes:

– cutting, shaping and finishing of stone for use in construction, in cemeteries, on roads, as roofing etc.
– manufacture of stone furniture

This class excludes:

- *activities carried out by operators of quarries, e.g. production of rough cut stone, see 08.11*
- *production of millstones, abrasive stones and similar products, see 23.9*

23.9   Manufacture of abrasive products and non-metallic mineral products n.e.c.

This group includes the manufacture of other non-metallic mineral products.

23.91   Production of abrasive products

This class includes:

- manufacture of millstones, sharpening or polishing stones and natural or artificial abrasive products on a support, including abrasive products on a soft base (e.g. sandpaper)

23.99   Manufacture of other non-metallic mineral products n.e.c.

This class includes:

- manufacture of friction material and unmounted articles thereof with a base of mineral substances or of cellulose
- manufacture of mineral insulating materials:
  - slag wool, rock wool and similar mineral wools; exfoliated vermiculite, expanded clays and similar heat-insulating, sound-insulating or sound-absorbing materials
- manufacture of articles of diverse mineral substances:
  - worked mica and articles of mica, of peat, of graphite (other than electrical articles) etc.
- manufacture of articles of asphalt or similar material, e.g. asphalt-based adhesives, coal tar pitch etc.
- manufacture of carbon and graphite fibres and products (except electrodes and electrical applications)
- manufacture of artificial corundum

This class excludes:

- *manufacture of glass wool and non-woven glass wool products, see 23.14*
- *manufacture of graphite electrodes, see 27.90*
- *manufacture of carbon or graphite gaskets, see 28.29*

24   Manufacture of basic metals

This division includes the activities of smelting and/or refining ferrous and non-ferrous metals from ore, pig or scrap, using electrometallurgic and other process metallurgic techniques. This division also includes the manufacture of metal alloys and super-alloys by introducing other chemical elements to pure metals. The output of smelting and refining, usually in ingot form, is used in rolling, drawing and extruding operations to make products such as plate, sheet, strip, bars, rods, wire or tubes, pipes and hollow profiles and in molten form to make castings and other basic metal products.

24.1   Manufacture of basic iron and steel and of ferro-alloys

This group includes activities such as direct reduction of iron ore, production of pig iron in molten or solid form, conversion of pig iron into steel, manufacture of ferro-alloys and manufacture of steel products.

24.10   Manufacture of basic iron and steel and of ferro-alloys

This class includes:

- operation of blast furnaces, steel converters, rolling and finishing mills
- production of pig iron and spiegeleisen in pigs, blocks or other primary forms
- production of ferro-alloys
- production of ferrous products by direct reduction of iron and other spongy ferrous products;
- production of iron of exceptional purity by electrolysis or other chemical processes
- remelting of scrap ingots of iron or steel
- production of granular iron and iron powder
- production of steel in ingots or other primary forms
- production of semi-finished products of steel
- manufacture of hot-rolled and cold-rolled flat-rolled products of steel
- manufacture of hot-rolled bars and rods of steel
- manufacture of hot-rolled open sections of steel
- manufacture of sheet piling of steel and welded open sections of steel
- manufacture of railway track materials (unassembled rails) of steel

This class excludes:

- *cold drawing of bars, see 24.31*

24.2   **Manufacture of tubes, pipes, hollow profiles and related fittings, of steel**

24.20   Manufacture of tubes, pipes, hollow profiles and related fittings, of steel

This class includes:

- manufacture of seamless tubes and pipes of circular or non-circular cross section and of blanks of circular cross section, for further processing, by hot rolling, hot extrusion or by other hot processes, of an intermediate product which can be a bar or a billet obtained by hot rolling or continuous casting
- manufacture of precision and non-precision seamless tubes and pipes from hot rolled or hot extruded blanks by further processing, by cold-drawing or cold-rolling of tubes and pipes of circular cross section and by cold drawing only for tubes and pipes of non circular cross section and hollow profiles
- manufacture of welded tubes and pipes of an external diameter exceeding 406.4 mm, cold formed from hot rolled flat products and longitudinally or spirally welded
- manufacture of welded tubes and pipes of an external diameter of 406.4 mm or less of circular cross section by continuous cold or hot forming of hot or cold rolled flat products and longitudinally or spirally welded and of non-circular cross section by hot or cold forming into shape from hot or cold rolled strip longitudinally welded
- manufacture of welded precision tubes and pipes of an external diameter of 406.4 mm or less by hot or cold forming of hot or cold rolled strip and longitudinally welded delivered as welded or further processed, by cold drawing or cold rolling or cold formed into shape for tube and pipe of non-circular cross section
- manufacture of flat flanges and flanges with forged collars by processing of hot rolled flat products of steel
- manufacture of butt-welding fittings, such as elbows and reductions, by forging of hot rolled seamless tubes of steel
- threaded and other tube or pipe fittings of steel

This class excludes:

- *manufacture of seamless tubes and pipes of steel by centrifugally casting, see 24.52*

24.3   **Manufacture of other products of first processing of steel**

This group includes manufacturing other products by cold processing of steel.

24.31   Cold drawing of bars

This class includes:

- manufacture of steel bars and solid sections of steel by cold drawing, grinding or turning

This class excludes:

- *drawing of wire, see 24.34*

24.32   Cold rolling of narrow strip

This class includes:

- manufacture of coated or uncoated flat rolled steel products in coils or in straight lengths of a width less than 600 mm by cold re-rolling of hot-rolled flat products or of steel rod

24.33   Cold forming or folding

This class includes:

- manufacture of open sections by progressive cold forming on a roll mill or folding on a press of flat-rolled products of steel
- manufacture of cold-formed or cold-folded, ribbed sheets and sandwich panels.

24.34   Cold drawing of wire

This class includes:

- manufacture of drawn steel wire, by cold drawing of steel wire rod

This class excludes:

- *drawing of bars and solid sections of steel, see 24.31*
- *manufacture of derived wire products, see 25.93*

24.4   **Manufacture of basic precious and other non-ferrous metals**

24.41   Precious metals production

This class includes:

- production of basic precious metals:
  - ■ production and refining of unwrought or wrought precious metals: gold, silver, platinum etc. from ore and scrap
- production of precious metal alloys
- production of precious metal semi-products
- production of silver rolled onto base metals

– production of gold rolled onto base metals or silver
– production of platinum and platinum group metals rolled onto gold, silver or base metals

This class also includes:

– manufacture of wire of these metals by drawing
– manufacture of precious metal foil laminates

This class excludes:

– *casting of non-ferrous metals, see 24.53, 24.54*
– *manufacture of precious metal jewellery, see 32.12*

24.42   Aluminium production

This class includes:

– production of aluminium from alumina
– production of aluminium from electrolytic refining of aluminium waste and scrap
– production of aluminium alloys
– semi-manufacturing of aluminium

This class also includes:

– manufacture of wire of these metals by drawing
– production of aluminium oxide (alumina)
– production of aluminium wrapping foil
– manufacture of aluminium foil laminates made from aluminium foil as primary component

This class excludes:

– *casting of non-ferrous metals, see 24.53, 24.54*

24.43   Lead, zinc and tin production

This class includes:

– production of lead, zinc and tin from ores
– production of lead, zinc and tin from electrolytic refining of lead, zinc and tin waste and scrap
– production of lead, zinc and tin alloys
– semi-manufacturing of lead, zinc and tin

This class also includes:

– manufacture of wire of these metals by drawing
– production of tin foil

This class excludes:

– *casting of non-ferrous metals, see 24.53, 24.54*

24.44   Copper production

This class includes:

– production of copper from ores
– production of copper from electrolytic refining of copper waste and scrap
– production of copper alloys
– manufacture of fuse wire or strip
– semi-manufacturing of copper

This class also includes:

– manufacture of wire of these metals by drawing

This class excludes:

– *casting of non-ferrous metals, see 24.53, 24.54*

24.45   Other non-ferrous metal production

This class includes:

– production of chrome, manganese, nickel etc. from ores or oxides
– production of chrome, manganese, nickel etc. from electrolytic and aluminothermic refining of chrome, manganese, nickel etc., waste and scrap

- production of alloys of chrome, manganese, nickel etc.
- semi-manufacturing of chrome, manganese, nickel etc.
- production of mattes of nickel

This class also includes:

- manufacture of wire of these metals by drawing

This class excludes:

- *casting of non-ferrous metals, see 24.53, 24.54*

24.46　Processing of nuclear fuel

This class includes:

- production of uranium metal from pitchblende or other ores
- smelting and refining of uranium

24.5　Casting of metals

This group includes the manufacture of semi-finished products and various castings by a casting process.

This group excludes:

- *manufacture of finished cast products such as:*
  - *boilers and radiators, see 25.21*
  - *cast household items, see 25.99*

24.51　Casting of iron

This class includes activities of iron foundries.

This class includes:

- casting of semi-finished iron products
- casting of grey iron castings
- casting of spheroidal graphite iron castings
- casting of malleable cast-iron products
- manufacture of tubes, pipes and hollow profiles and of tube or pipe fittings of cast-iron

24.52　Casting of steel

This class includes activities of steel foundries.

This class includes:

- casting of semi-finished steel products
- casting of steel castings
- manufacture of seamless tubes and pipes of steel by centrifugal casting
- manufacture of tube or pipe fittings of cast-steel

24.53　Casting of light metals

This class includes:

- casting of semi-finished products of aluminium, magnesium, titanium, zinc etc.
- casting of light metal castings

24.54　Casting of other non-ferrous metals

This class includes:

- casting of heavy metal castings
- casting of precious metal castings
- die-casting of non-ferrous metal castings

25　Manufacture of fabricated metal products, except machinery and equipment

This division includes the manufacture of "pure" metal products (such as parts, containers and structures), usually with a static, immovable function, as opposed to the following divisions 26-30, which cover the manufacture of combinations or assemblies of such metal products (sometimes with other materials) into more complex units that, unless they are purely electrical, electronic or optical, work with moving parts.

The manufacture of weapons and ammunition is also included in this division.

This division excludes:

- *specialised repair and maintenance activities, see 33.1*
- *specialised installation of manufactured goods produced in this division in buildings, such as central heating boilers, see 43.22*

25.1   **Manufacture of structural metal products**

This group includes the manufacture of structural metal products (such as metal frameworks or parts for construction).

25.11   **Manufacture of metal structures and parts of structures**

This class includes:

- manufacture of metal frameworks or skeletons for construction and parts thereof (towers, masts, trusses, bridges etc.)
- manufacture of industrial frameworks in metal (frameworks for blast furnaces, lifting and handling equipment etc.)
- manufacture of prefabricated buildings mainly of metal:
  - site huts, modular exhibition elements etc.

This class excludes:

- manufacture of parts for marine or power boilers, see 25.30
- manufacture of assembled railway track fixtures, see 25.99
- manufacture of sections of ships, see 30.11

25.12   **Manufacture of doors and windows of metal**

This class includes:

- manufacture of metal doors, windows and their frames, shutters and gates
- metal room partitions for floor attachment

25.2   **Manufacture of tanks, reservoirs and containers of metal**

This group includes the manufacture of tanks, central heating radiators and boilers.

25.21   **Manufacture of central heating radiators and boilers**

This class excludes:

- *manufacture of electrical ovens and water heaters, see 27.51*

25.29   **Manufacture of other tanks, reservoirs and containers of metal**

This class includes:

- manufacture of reservoirs, tanks and similar containers of metal, of types normally installed as fixtures for storage or manufacturing use
- manufacture of metal containers for compressed or liquefied gas

This class excludes:

- *manufacture of metal casks, drums, cans, pails, boxes etc. of a kind normally used for carrying and packing of goods of a capacity not exceeding 300 litres, see 25.91, 25.92*
- *manufacture of transport containers, see 29.20*
- *manufacture of tanks (armoured military vehicles), see 30.40*

25.3   **Manufacture of steam generators, except central heating hot water boilers**

This group includes the manufacture of steam generators.

25.30   **Manufacture of steam generators, except central heating hot water boilers**

This class includes:

- manufacture of steam or other vapour generators
- manufacture of auxiliary plant for use with steam generators:
  - condensers, economisers, superheaters, steam collectors and accumulators
- manufacture of nuclear reactors, except isotope separators
- manufacture of parts for marine or power boilers

This class also includes:

- pipe system construction comprising further processing of tubes generally to make pressure pipes or pipe systems together with the associated design and construction work

This class excludes:

- *manufacture of central heating hot-water boilers and radiators, see 25.21*

- *manufacture of boiler-turbine sets, see 28.11*
- *manufacture of isotope separators, see 28.99*

25.4　　Manufacture of weapons and ammunition

25.40　　Manufacture of weapons and ammunition

This class includes:

- manufacture of heavy weapons (artillery, mobile guns, rocket launchers, torpedo tubes, heavy machine guns)
- manufacture of small arms (revolvers, shotguns, light machine guns)
- manufacture of air or gas guns and pistols
- manufacture of war ammunition

This class also includes:

- manufacture of hunting, sporting or protective firearms and ammunition
- manufacture of explosive devices such as bombs, mines and torpedoes

This class excludes:

- *manufacture of percussion caps, detonators or signalling flares, see 20.51*
- *manufacture of cutlasses, swords, bayonets etc., see 25.71*
- *manufacture of armoured vehicles for the transport of banknotes or valuables, see 29.10*
- *manufacture of space vehicles, see 30.30*
- *manufacture of tanks and other fighting vehicles, see 30.40*
- *manufacture of intercontinental ballistic missiles (ICBM), see 30.30*

25.5　　Forging, pressing, stamping and roll-forming of metal; powder metallurgy

This group includes general activities for the treatment of metal, such as forging or pressing, which are typically carried out on a fee or contract basis.

25.50　　Forging, pressing, stamping and roll-forming of metal; powder metallurgy

This class includes:

- forging, pressing, stamping and roll-forming of metal
- powder metallurgy: production of metal objects directly from metal powders by heat treatment (sintering) or under pressure

This class excludes:

- *production of metal powder, see 24.1, 24.4*

25.6　　Treatment and coating of metals; machining

This group includes general activities for the treatment of metal, such as plating, coating, engraving, boring, polishing, welding etc., which are typically carried out on a fee or contract basis.

25.61　　Treatment and coating of metals

This class includes:

- plating, anodising etc. of metals
- heat treatment of metals
- deburring, sandblasting, tumbling, cleaning of metals
- colouring, engraving, of metals
- non-metallic coating of metals:
  - plasticising, enamelling, lacquering etc.
- hardening, buffing of metals

This class excludes:

- *activities of farriers, see 01.62*
- *printing onto metals, see 18.12*
- *metal coating of plastics, see 22.29*
- *rolling precious metals onto base metals or other metals, see 24.41, 24.42, 24.43, 24.44*
- *"while-you-wait" engraving services, see 95.29*

25.62　　Machining

This class includes:

- boring, turning, milling, eroding, planing, lapping, broaching, levelling, sawing, grinding, sharpening, polishing, welding, splicing etc. of metalwork pieces
- cutting of and writing on metals by means of laser beams

This class excludes:

– *activities of farriers, see 01.62*

25.7 Manufacture of cutlery, tools and general hardware

This group includes the manufacture of cutlery; metal hand tools and general hardware.

25.71 Manufacture of cutlery

This class includes:

– manufacture of domestic cutlery such as knives, forks, spoons etc.
– manufacture of other articles of cutlery:
  - cleavers and choppers
  - razors and razor blades
  - scissors and hair clippers
– manufacture of cutlasses, swords, bayonets etc.

This class excludes:

– *manufacture of hollowware (pots, kettles etc.), dinnerware (bowls, platters etc.) or flatware (plates, saucers etc.), see 25.99*
– *manufacture of cutlery of precious metal, see 32.12*

25.72 Manufacture of locks and hinges

This class includes:

– manufacture of padlocks, locks, keys, hinges and the like, hardware for buildings, furniture, vehicles etc.

25.73 Manufacture of tools

This class includes:

– manufacture of knives and cutting blades for machines or for mechanical appliances
– manufacture of hand tools such as pliers, screwdrivers etc.
– manufacture of non-power-driven agricultural hand tools
– manufacture of saws and saw blades, including circular saw blades and chainsaw blades
– manufacture of interchangeable tools for hand tools, whether or not power-operated, or for machine tools: drills, punches, milling cutters etc.
– manufacture of press tools
– manufacture of blacksmiths' tools: forges, anvils etc.
– manufacture of moulding boxes and moulds (except ingot moulds)
– manufacture of vices, clamps

This class excludes:

– *manufacture of power-driven hand tools, see 28.24*
– *manufacture of ingot moulds, see 28.91*

25.9 Manufacture of other fabricated metal products

This group includes the manufacture of a variety of metal products, such as cans and buckets; nails, bolts and nuts; metal household articles; metal fixtures; ships propellers and anchors; assembled railway track fixtures etc. for a variety of household and industrial uses.

25.91 Manufacture of steel drums and similar containers

This class includes:

– manufacture of pails, cans, drums, buckets, boxes

This class excludes:

– *manufacture of tanks and reservoirs, see 25.2*

25.92 Manufacture of light metal packaging

This class includes:

– manufacture of tins and cans for food products, collapsible tubes and boxes
– manufacture of metallic closures

25.93 Manufacture of wire products, chain and springs

This class includes:

– manufacture of metal cable, plaited bands and similar articles

– manufacture of uninsulated metal cable or insulated cable not capable of being used as a conductor of electricity
– manufacture of coated or cored wire
– manufacture of articles made of wire: barbed wire, wire fencing, grill, netting, cloth etc.
– coated electrodes for electric arc-welding
– manufacture of nails and pins
– manufacture of springs (except watch springs):
  ■ leaf springs, helical springs, torsion bar springs
  ■ leaves for springs
– manufacture of chain, except power transmission chain

This class excludes:

– *manufacture of clock or watch springs, see 26.52*
– *manufacture of wire and cable for electricity transmission, see 27.32*
– *manufacture of power transmission chain, see 28.15*

25.94   Manufacture of fasteners and screw machine products

This class includes:

– manufacture of rivets, washers and similar non-threaded products
– manufacture of screw machine products
– manufacture of bolts, screws, nuts and similar threaded products

25.99   Manufacture of other fabricated metal products n.e.c.

This class includes:

– manufacture of metal household articles:
  ■ flatware: plates, saucers etc.
  ■ hollowware: pots, kettles etc.
  ■ dinnerware: bowls, platters etc.
  ■ saucepans, frying pans and other non-electrical utensils for use at the table or in the kitchen
  ■ small hand-operated kitchen appliances and accessories
  ■ metal scouring pads
– manufacture of building components of zinc: gutters, roof capping, baths, sinks, washbasins and similar articles
– manufacture of metal goods for office use, except furniture
– manufacture of safes, strongboxes, armoured doors etc.
– manufacture of various metal articles:
  ■ ship propellers and blades thereof
  ■ anchors
  ■ bells
  ■ assembled railway track fixtures
  ■ clasps, buckles, hooks
  ■ metal ladder
  ■ metal signs, including road signs
– manufacture of foil bags
– manufacture of permanent metallic magnets
– manufacture of metal vacuum jugs and bottles
– manufacture of metal badges and metal military insignia
– manufacture of metal hair curlers, metal umbrella handles and frames, combs

This class excludes:

– *manufacture of swords, bayonets, see 25.71*
– *manufacture of shopping carts, see 30.99*
– *manufacture of metal furniture, see 31.01, 31.02, 31.09*
– *manufacture of sports goods, see 32.30*
– *manufacture of games and toys, see 32.40*

26   Manufacture of computer, electronic and optical products

This division includes the manufacture of computers, computer peripherals, communications equipment, and similar electronic products, as well as the manufacture of components for such products. Production processes of this division are characterised by the design and use of integrated circuits and the application of highly specialised miniaturisation technologies.

C

The division also contains the manufacture of consumer electronics, measuring, testing and navigating equipment, irradiation, electromedical and electrotherapeutic equipment, optical instruments and equipment, and the manufacture of magnetic and optical media.

26.1   **Manufacture of electronic components and boards**

26.11   Manufacture of electronic components

This class includes the manufacture of semiconductors and other components for electronic applications.

This class includes:

- manufacture of capacitors, electronic
- manufacture of resistors, electronic
- manufacture of microprocessors
- manufacture of electron tubes
- manufacture of electronic connectors
- manufacture of bare printed circuit boards
- manufacture of integrated circuits (analogue, digital or hybrid)
- manufacture of diodes, transistors, and related discrete devices
- manufacture of inductors (e.g. chokes, coils, transformers), electronic component type
- manufacture of electronic crystals and crystal assemblies
- manufacture of solenoids, switches, and transducers for electronic applications
- manufacture of dice or wafers, semiconductor, finished or semi-finished
- manufacture of display components (plasma, polymer, LCD)
- manufacture of light emitting diodes (LED)

This class also includes:

- manufacture of printer cables, monitor cables, USB cables, connectors etc.

This class excludes:

- *printing of smart cards, see 18.12*
- *manufacture of computer and television displays, see 26.20, 26.40*
- *manufacture of modems (carrier equipment), see 26.30*
- *manufacture of X-ray tubes and similar irradiation devices, see 26.60*
- *manufacture of optical equipment and instruments, see 26.70*
- *manufacture of similar devices for electrical applications, see division 27*
- *manufacture of fluorescent ballasts, see 27.11*
- *manufacture of electrical relays, see 27.12*
- *manufacture of electrical wiring devices, see 27.33*
- *manufacture of complete equipment is classified elsewhere based on complete equipment classification*

26.12   Manufacture of loaded electronic boards

This class includes:

- manufacture of loaded printed circuit boards
- loading of components onto printed circuit boards
- manufacture of interface cards (e.g. sound, video, controllers, network, modems)

This class excludes:

- *printing of smart cards, see 18.12*
- *manufacture of bare printed circuit boards, see 26.11*

26.2   **Manufacture of computers and peripheral equipment**

26.20   Manufacture of computers and peripheral equipment

This class includes the manufacture and/or assembly of electronic computers, such as mainframes, desktop computers, laptops and computer servers; and computer peripheral equipment, such as storage devices and input/output devices (printers, monitors, keyboards). Computers can be analog, digital, or hybrid. Digital computers, the most common type, are devices that do all of the following: (1) store the processing program or programs and the data immediately necessary for the execution of the program; (2) can be freely programmed in accordance with the requirements of the user; (3) perform arithmetical computations specified by the user; and (4) execute, without human intervention, a processing program that requires the computer to modify its execution by logical decision during the processing run. Analog computers are capable of simulating mathematical models and comprise at least analog control and programming elements.

This class includes:

- manufacture of desktop computers
- manufacture of laptop computers
- manufacture of main frame computers
- manufacture of hand-held computers (e.g. PDA)
- manufacture of magnetic disk drives, flash drives and other storage devices
- manufacture of optical (e.g. CD-RW, CD-ROM, DVD-ROM, DVD-RW) disk drives
- manufacture of printers
- manufacture of monitors
- manufacture of keyboards
- manufacture of all types of mice, joysticks, and trackball accessories
- manufacture of dedicated computer terminals
- manufacture of computer servers
- manufacture of scanners, including bar code scanners
- manufacture of smart card readers
- manufacture of virtual reality helmets
- manufacture of computer projectors (video beamers)

This class also includes:

- manufacture of computer terminals, like automatic teller machines (ATMs), point-of-sale (POS) terminals, not mechanically operated
- manufacture of multi-function office equipment performing two or more of following functions: printing, scanning, copying, faxing

This class excludes:

- *reproduction of recorded media (computer media, sound, video, etc.), see 18.20*
- *manufacture of electronic components and electronic assemblies used in computers and peripherals, see 26.1*
- *manufacture of internal/external computer modems, see 26.12*
- *manufacture of interface cards, modules and assemblies, see 26.12*
- *manufacture of loaded electronic boards, see 26.12*
- *manufacture of modems, carrier equipment, see 26.30*
- *manufacture of digital communication switches, data communications equipment (e.g. bridges, routers, gateways), see 26.30*
- *manufacture of consumer electronic devices, such as CD players and DVD players, see 26.40*
- *manufacture of television monitors and displays, see 26.40*
- *manufacture of video game consoles, see 26.40*
- *manufacture of blank optical and magnetic media for use with computers or other devices, see 26.80*

26.3　　　　　　**Manufacture of communication equipment**

26.30　　　　　**Manufacture of communication equipment**

26.30/1　　　　**Manufacture of telegraph and telephone apparatus and equipment**

This subclass includes:

- manufacture of central office switching equipment
- manufacture of cordless telephones
- manufacture of private branch exchange (PBX) equipment
- manufacture of telephone and facsimile equipment, including telephone answering machines
- manufacture of data communications equipment, such as bridges, routers, and gateways
- manufacture of pagers
- manufacture of cellular phones
- manufacture of mobile communication equipment
- manufacture of modems, carrier equipment
- manufacture of burglar and fire alarm systems, sending signals to a control station
- manufacture of communication devices using infrared signal (e.g. remote controls)

This subclass excludes:

- *manufacture of electronic components and subassemblies used in communications equipment, including internal/external computer modems, see 26.1*
- *manufacture of loaded electronic boards, see 26.12*
- *manufacture of computers and computer peripheral equipment, see 26.20*

C

– *manufacture of consumer audio and video equipment, see 26.40*

– *manufacture of GPS devices, see 26.51*

– *manufacture of electronic scoreboards, see 27.90*

– *manufacture of traffic lights, see 27.90*

26.30/9 Manufacture of communication equipment (other than telegraph and telephone apparatus and equipment)

This subclass includes:

– manufacture of transmitting and receiving antenna

– manufacture of cable television equipment

– manufacture of radio and television studio and broadcasting equipment, including television cameras

– manufacture of radio and television transmitters

This subclass excludes:

– *manufacture of electronic components and subassemblies used in communications equipment, including internal/external computer modems, see 26.1*

– *manufacture of loaded electronic boards, see 26.12*

– *manufacture of computers and computer peripheral equipment, see 26.20*

– *manufacture of consumer audio and video equipment, see 26.40*

– *manufacture of GPS devices, see 26.51*

– *manufacture of electronic scoreboards, see 27.90*

– *manufacture of traffic lights, see 27.90*

26.4 **Manufacture of consumer electronics**

26.40 Manufacture of consumer electronics

This class includes the manufacture of electronic audio and video equipment for home entertainment, motor vehicle, public address systems and musical instrument amplification.

This class includes:

– manufacture of video cassette recorders and duplicating equipment

– manufacture of televisions

– manufacture of television monitors and displays

– manufacture of audio recording and duplicating systems

– manufacture of stereo equipment

– manufacture of radio receivers

– manufacture of speaker systems

– manufacture of household-type video cameras

– manufacture of jukeboxes

– manufacture of amplifiers for musical instruments and public address systems

– manufacture of microphones

– manufacture of CD and DVD players

– manufacture of karaoke machines

– manufacture of headphones (e.g. radio, stereo, computer)

– manufacture of video game consoles

This class excludes:

– *reproduction of recorded media (computer media, sound, video, etc.), see 18.2*

– *manufacture of computer peripheral devices and computer monitors, see 26.20*

– *manufacture of telephone answering machines, see 26.30*

– *manufacture of paging equipment, see 26.30*

– *manufacture of remote control devices (radio and infrared), see 26.30*

– *manufacture of broadcast studio equipment such as reproduction equipment, transmitting and receiving antennas, commercial video cameras, see 26.30*

– *manufacture of antennas, see 26.30*

– *manufacture of digital cameras, see 26.70*

– *manufacture of electronic games with fixed (non-replaceable) software, see 32.40*

26.5 **Manufacture of instruments and appliances for measuring, testing and navigation; watches and clocks**

This group includes the manufacture of measuring, testing and navigating equipment for various industrial and non-industrial purposes, including time-based measuring devices such as watches and clocks and related devices.

26.51　　**Manufacture of instruments and appliances for measuring, testing and navigation**

This class comprises manufacturing of search, detection, navigation, guidance, aeronautical, and nautical systems and instruments; automatic controls and regulators for applications, such as heating, air conditioning, refrigeration and appliances; instruments and devices for measuring, displaying, indicating, recording, transmitting, and controlling temperature, humidity, pressure, vacuum, combustion, flow, level, viscosity, density, acidity, concentration, and rotation; totalising (i.e., registering) fluid meters and counting devices; instruments for measuring and testing the characteristics of electricity and electrical signals; instruments and instrumentation systems for laboratory analysis of the chemical or physical composition or concentration of samples of solid, fluid, gaseous, or composite material; other measuring and testing instruments and parts thereof.

The manufacture of non-electric measuring, testing and navigating equipment (except simple mechanical tools) is included here.

This class includes:

- manufacture of aircraft engine instruments
- manufacture of automotive emissions testing equipment
- manufacture of meteorological instruments
- manufacture of physical properties testing and inspection equipment
- manufacture of polygraph machines
- manufacture of radiation detection and monitoring instruments
- manufacture of surveying instruments
- manufacture of thermometers liquid-in-glass and bimetal types (except medical)
- manufacture of humidistats
- manufacture of hydronic limit controls
- manufacture of flame and burner control
- manufacture of spectrometers
- manufacture of pneumatic gauges
- manufacture of consumption meters (e.g., water, gas, electricity)
- manufacture of flow meters and counting devices
- manufacture of tally counters
- manufacture of mine detectors, pulse (signal) generators; metal detectors
- manufacture of search, detection, navigation, aeronautical, and nautical equipment, including sonobuoys
- manufacture of radar equipment
- manufacture of GPS devices
- manufacture of environmental controls and automatic controls for appliances
- manufacture of measuring and recording equipment (e.g. flight recorders)
- manufacture of motion detectors
- manufacture of radars
- manufacture of laboratory analytical instruments (e.g. blood analysis equipment)
- manufacture of laboratory scales, balances, incubators, and miscellaneous laboratory apparatus for measuring, testing, etc.
- manufacture of industrial process control equipment

This class excludes:

- *manufacture of telephone answering machines, see 26.30*
- *manufacture of irradiation equipment, see 26.60*
- *manufacture of optical positioning equipment, see 26.70*
- *manufacture of dictating machines, see 28.23*
- *manufacture of weighing devices (other than laboratory balances), levels, tape measures etc., see 28.29*
- *manufacture of medical thermometers, see 32.50*
- *installation of industrial process control equipment see 33.20*
- *manufacture of simple mechanical measuring tools (e.g. measuring tapes, calipers), see manufacturing class according to main material used*

26.51/1　　**Manufacture of electronic instruments and appliances for measuring, testing, and navigation, except industrial process control equipment**

26.51/2　　**Manufacture of electronic industrial process control equipment**

This subclass includes:

- manufacture of components for electronic industrial continuous process control systems (also for automated production plants consisting of various machines, handling devices and centralised controlling apparatus)

26.51/3　　**Manufacture of non-electronic instruments and appliances for measuring, testing and navigation, except industrial process control equipment**

26.51/4   Manufacture of non-electronic industrial process control equipment

This subclass includes:

- manufacture of components for non-electronic industrial continuous process control systems (also for automated production plants consisting of various machines, handling devices and centralised controlling apparatus)

26.52   Manufacture of watches and clocks

This class includes the manufacture of watches, clocks and timing mechanisms and parts thereof.

This class includes:

- manufacture of watches and clocks of all kinds, including instrument panel clocks
- manufacture of watch and clock cases, including cases of precious metals
- manufacture of time-recording equipment and equipment for measuring, recording and otherwise displaying intervals of time with a watch or clock movement or with synchronous motor, such as:
  - parking meters
  - time clocks
  - time/date stamps
  - process timers
- manufacture of time switches and other releases with a watch or clock movement or with synchronous motor:
  - time locks
- manufacture of components for clocks and watches:
  - movements of all kinds for watches and clocks
  - springs, jewels, dials, hands, plates, bridges and other parts
  - watch and clock cases and housings of all materials

This class excludes:

- *manufacture of non-metal watch bands (textile, leather, plastic), see 15.12*
- *manufacture of watch bands of precious metal, see 32.12*
- *manufacture of watch bands of non-precious metal, see 32.13*

26.6   **Manufacture of irradiation, electromedical and electrotherapeutic equipment**

26.60   Manufacture of irradiation, electromedical and electrotherapeutic equipment

This class includes:

- manufacture of irradiation apparatus and tubes (e.g. industrial, medical diagnostic, medical therapeutic, research, scientific):
  - beta-, gamma, X-ray or other radiation equipment
- manufacture of CT scanners
- manufacture of PET scanners
- manufacture of magnetic resonance imaging (MRI) equipment
- manufacture of medical ultrasound equipment
- manufacture of electrocardiographs
- manufacture of electromedical endoscopic equipment
- manufacture of medical laser equipment
- manufacture of pacemakers
- manufacture of hearing aids

This class also includes:

- manufacture of food and milk irradiation equipment

This class excludes:

- *manufacture of tanning beds, see 27.90*

26.7   **Manufacture of optical instruments and photographic equipment**

26.70   Manufacture of optical instruments and photographic equipment

This class includes the manufacture of optical instruments and lenses, and the manufacture of photographic equipment.

This class excludes:

- *manufacture of complete equipment using laser components, see manufacturing class by type of machinery (e.g. medical laser equipment, see 26.60)*
- *manufacture of ophthalmic goods, see 32.50*
- *manufacture of computer projectors, see 26.20*

- *manufacture of commercial TV and video cameras, see 26.30*
- *manufacture of household-type video cameras, see 26.40*
- *manufacture of photocopy machinery, see 28.23*

**26.70/1	Manufacture of optical precision instruments**

This subclass includes the manufacture of optical instruments and lenses, such as binoculars, microscopes (except electron, proton), telescopes, prisms, and lenses (except ophthalmic); the coating or polishing of lenses (except ophthalmic) and the mounting of lenses (except ophthalmic).

This subclass includes:

- manufacture of optical mirrors
- manufacture of optical gun sighting equipment
- manufacture of optical positioning equipment
- manufacture of optical magnifying instruments
- manufacture of optical machinist's precision tools
- manufacture of optical comparators
- manufacture of optical measuring and checking devices and instruments (e.g. fire control equipment, range finders)
- manufacture of lenses (except ophthalmic), optical microscopes, binoculars and telescopes
- manufacture of laser assemblies

**26.70/2	Manufacture of photographic and cinematographic equipment**

This subclass includes the manufacture of photographic equipment such as cameras and light meters.

This subclass includes:

- manufacture of cameras (film and digital)
- manufacture of motion picture and slide projectors
- manufacture of overhead transparency projectors

**26.8	Manufacture of magnetic and optical media**

**26.80	Manufacture of magnetic and optical media**

This class includes the manufacture of magnetic and optical recording media,

This class includes:

- manufacture of blank magnetic audio and video tapes
- manufacture of blank magnetic audio and video cassettes
- manufacture of blank diskettes
- manufacture of blank optical discs
- manufacture of hard drive media

This class excludes:

- *reproduction of recorded media (computer media, sound, video, etc.), see 18.2*

**27	Manufacture of electrical equipment**

This division includes the manufacture of products that generate, distribute and use electrical power. Also included is the manufacture of electrical lighting, signalling equipment and electric household appliances.

This division excludes the manufacture of electronic products (see division 26).

**27.1	Manufacture of electric motors, generators, transformers and electricity distribution and control apparatus**

This group comprises the manufacture of power, distribution and specialty transformers; electric motors, generators, and motor generator sets.

**27.11	Manufacture of electric motors, generators and transformers**

This class includes manufacture of all electric motors and transformers: AC, DC and AC/DC.

This class includes:

- manufacture of electric motors (except internal combustion engine starting motors)
- manufacture of distribution transformers, electric
- manufacture of arc-welding transformers
- manufacture of fluorescent ballasts (i.e. transformers)
- manufacture of substation transformers for electric power distribution

C

- manufacture of transmission and distribution voltage regulators
- manufacture of power generators (except battery charging alternators for internal combustion engines)
- manufacture of motor generator sets (except turbine generator set units)
- rewinding of armatures on a factory basis

This class excludes:

- *manufacture of electronic component-type transformers and switches, see 26.11*
- *manufacture of electric welding and soldering equipment, see 27.90*
- *manufacture of solid state inverters, rectifiers and converters, see 27.90*
- *manufacture of turbine-generator sets, see 28.11*
- *manufacture of starting motors and generators for internal combustion engines, see 29.31*

27.12   Manufacture of electricity distribution and control apparatus

This class includes:

- manufacture of power circuit breakers
- manufacture of surge suppressors (for distribution level voltage)
- manufacture of control panels for electric power distribution
- manufacture of electrical relays
- manufacture of duct for electrical switchboard apparatus
- manufacture of electric fuses
- manufacture of power switching equipment
- manufacture of electric power switches (except pushbutton, snap, solenoid, tumbler)
- manufacture of prime mover generator sets

This class excludes:

- *manufacture of environmental controls and industrial process control instruments, see 26.51*
- *manufacture of switches for electrical circuits, such as pushbutton and snap switches, see 27.33*

27.2   Manufacture of batteries and accumulators

27.20   Manufacture of batteries and accumulators

This class includes the manufacture of non-rechargeable and rechargeable batteries.

This class includes:

- manufacture of primary cells and primary batteries
  - cells containing manganese dioxide, mercuric dioxide, silver oxide etc.
- manufacture of electric accumulators, including parts thereof:
  - separators, containers, covers
- manufacture of lead acid batteries
- manufacture of NiCad batteries
- manufacture of NiMH batteries
- manufacture of lithium batteries
- manufacture of dry cell batteries
- manufacture of wet cell batteries

27.3   Manufacture of wiring and wiring devices

This group includes the manufacture of current-carrying wiring devices and non current-carrying wiring devices for wiring electrical circuits regardless of material. This group also includes the insulating of wire and the manufacture of fibre optic cables.

27.31   Manufacture of fibre optic cables

This class includes:

- manufacture of fibre optic cable for data transmission or live transmission of images

This class excludes:

- *manufacture of glass fibres or strand, see 23.14*
- *manufacture of optical cable sets or assemblies with connectors or other attachments, see depending on application, e.g. 26.11*

27.32   Manufacture of other electronic and electric wires and cables

This class includes:

- manufacture of insulated wire and cable, made of steel, copper, aluminium

C

This class excludes:

- *manufacture (drawing) of wire, see 24.34, 24.41, 24.42, 24.43, 24.44, 24.45*
- *manufacture of computer cables, printer cables, USB cables and similar cable sets or assemblies, see 26.11*
- *manufacture of electrical cord sets with insulated wire and connectors, see 27.90*
- *manufacture of cable sets, wiring harnesses and similar cable sets or assemblies for automotive applications, see 29.31*

27.33　Manufacture of wiring devices

This class includes the manufacture of current-carrying and non current-carrying wiring devices for electrical circuits regardless of material.

This class includes:

- manufacture of bus bars, electrical conductors (except switchgear-type)
- manufacture of GFCI (ground fault circuit interrupters)
- manufacture of lamp holders
- manufacture of lightning arrestors and coils
- manufacture of switches for electrical wiring (e.g. pressure, pushbutton, snap, tumbler switches)
- manufacture of electrical outlets or sockets
- manufacture of boxes for electrical wiring (e.g. junction, outlet, switch boxes)
- manufacture of electrical conduit and fitting
- manufacture of transmission pole and line hardware
- manufacture of plastic non-current carrying wiring devices including plastic junction boxes, face plates and similar, plastic pole line fittings and switch covers

This class excludes:

- *manufacture of ceramic insulators, see 23.43*
- *manufacture of electronic component-type connectors, sockets, and switches, see 26.11*

27.4　Manufacture of electric lighting equipment

27.40　Manufacture of electric lighting equipment

This class includes the manufacture of electric light bulbs and tubes, and parts and components thereof (except glass blanks for electric light bulbs), electric lighting fixtures and lighting fixture components (except current-carrying devices).

This class includes:

- manufacture of discharge, incandescent, fluorescent, ultra-violet, infra-red etc. lamps, fixtures and bulbs
- manufacture of ceiling lighting fixtures
- manufacture of chandeliers
- manufacture of table lamps (i.e. lighting fixture)
- manufacture of Christmas tree lighting sets
- manufacture of electric fireplace logs
- manufacture of flashlights
- manufacture of electric insect lamps
- manufacture of lanterns (e.g. carbide, electric, gas, gasoline, kerosene)
- manufacture of spotlights
- manufacture of street lighting fixtures (except traffic signals)
- manufacture of lighting equipment for transportation equipment (e.g. for motor vehicles, aircraft, boats)

This class also includes:

- manufacture of non-electrical lighting equipment

This class excludes:

- *manufacture of glassware and glass parts for lighting fixtures, see 23.19*
- *manufacture of current-carrying wiring devices for lighting fixtures, see 27.33*
- *manufacture of ceiling fans or bath fans with integrated lighting fixtures, see 27.51*
- *manufacture of electrical signalling equipment such as traffic lights and pedestrian signalling equipment, see 27.90*
- *manufacture of electrical signs, see 27.90*

27.5　Manufacture of domestic appliances

This group includes the manufacture of small electric appliances and electric housewares, household-type fans, household-type vacuum cleaners, electric household-type floor care machines, household-type cooking appliances, household-type laundry equipment, household-type refrigerators, upright and chest freezers, and other electrical and non-electrical household appliances, such as dishwashers, water heaters, and garbage disposal units. This group includes the manufacture of appliances with electric, gas or other fuel sources.

27.51 **Manufacture of electric domestic appliances**

This class includes:

– manufacture of domestic electric appliances:
  - refrigerators
  - freezers
  - dishwashers
  - washing and drying machines
  - vacuum cleaners
  - floor polishers
  - waste disposers
  - grinders, blenders, juice squeezers
  - tin openers
  - electric shavers, electric toothbrushes, and other electric personal care device
  - knife sharpeners
  - ventilating or recycling hoods
– manufacture of domestic electrothermic appliances:
  - electric water heaters
  - electric blankets
  - electric dryers, combs, brushes, curlers
  - electric smoothing irons
  - space heaters and household-type fans, portable
  - electric ovens
  - microwave ovens
  - cookers, hotplates
  - toasters
  - coffee or tea makers
  - frying pans, roasters, grills, hoods
  - electric heating resistors etc.

This class excludes:

– *manufacture of commercial and industrial refrigerators and freezers, room air-conditioners, attic fans, permanently mounted space heaters, and commercial ventilation and exhaust fans, commercial-type cooking equipment; commercial-type laundry, dry-cleaning, and pressing equipment; commercial, industrial, and institutional vacuum cleaners, see division 28*
– *manufacture of household-type sewing machines, see 28.94*
– *installation of central vacuum cleaning systems, 43.29*

27.52 **Manufacture of non-electric domestic appliances**

This class includes:

– manufacture of domestic non-electric cooking and heating equipment:
  - non-electric space heaters, cooking ranges, grates, stoves, water heaters, cooking appliances, plate warmers

27.9 **Manufacture of other electrical equipment**

27.90 Manufacture of other electrical equipment

This class includes the manufacture of miscellaneous electrical equipment other than motors, generators and transformers, batteries and accumulators, wires and wiring devices, lighting equipment or domestic appliances.

This class includes:

– manufacture of battery chargers, solid-state
– manufacture of door opening and closing devices, electrical
– manufacture of electric bells
– manufacture of extension cords made from purchased insulated wire
– manufacture of ultrasonic cleaning machines (except laboratory and dental)
– manufacture of solid state inverters, rectifying apparatus, fuel cells, regulated and unregulated power supplies
– manufacture of uninterruptible power supplies (UPS)
– manufacture of surge suppressors (except for distribution level voltage)
– manufacture of appliance cords, extension cords, and other electrical cord sets with insulated wire and connectors

- manufacture of carbon and graphite electrodes, contacts, and other electrical carbon and graphite products
- manufacture of particle accelerators
- manufacture of electrical capacitors, resistors, condensers, and similar components
- manufacture of electromagnets
- manufacture of sirens
- manufacture of electronic scoreboards
- manufacture of electrical signs
- manufacture of electrical signalling equipment such as traffic lights and pedestrian signalling equipment
- manufacture of electrical insulators (except glass or porcelain), base metal conduit and fittings
- manufacture of electrical welding and soldering equipment, including hand-held soldering irons

This class excludes:

- *manufacture of porcelain electrical insulators, see 23.43*
- *manufacture of carbon and graphite fibres and products (except electrodes and electrical applications), see 23.99*
- *manufacture of electronic component-type rectifiers, voltage regulating integrated circuits, power converting integrated circuits, electronic capacitors, electronic resistors, and similar devices, see 26.11*
- *manufacture of transformers, motors, generators, switchgear, relays, and industrial controls, see 27.1*
- *manufacture of batteries, see 27.20*
- *manufacture of communication and energy wire, current-carrying and non current-carrying wiring devices, see 27.3*
- *manufacture of lighting equipment, see 27.40*
- *manufacture of household-type appliances, see 27.5*
- *manufacture of non-electrical welding and soldering equipment, see 28.29*
- *manufacture of motor vehicle electrical equipment, such as generators, alternators, spark plugs, ignition wiring harnesses, power window and door systems, voltage regulators, see 29.31*

28   Manufacture of machinery and equipment n.e.c.

This division includes the manufacture of machinery and equipment that acts independently on materials either mechanically or thermally or performs operations on materials (such as handling, spraying, weighing or packing), It includes their mechanical components which produce and apply force, and any specially manufactured primary parts. This includes the manufacture of fixed and mobile or hand-held devices, regardless of whether they are designed for industrial, building and civil engineering, agricultural or home use. The manufacture of special equipment for passenger or freight transport within demarcated premises also belongs within this division.

This division distinguishes between the manufacture of special-purpose machinery, i.e. machinery for exclusive use in a specific industry or a small cluster of specific industries, and general-purpose machinery, i.e. machinery that is being used in a wide range of industries.

This division also includes the manufacture of other special purpose machinery, not covered elsewhere in the classification, whether or not used in a manufacturing process, such as fairground amusement equipment, automatic bowling alley equipment, etc.

This division excludes the manufacture of metal products for general use (division 25), associated control devices, computer equipment, measurement and testing equipment, electricity distribution and control apparatus (divisions 26 and 27) and general-purpose motor vehicles (divisions 29 and 30).

28.1   Manufacture of general purpose machinery

28.11   Manufacture of engines and turbines, except aircraft, vehicle and cycle engines

This class includes:

- manufacture of internal combustion piston engines, except motor vehicle, aircraft and cycle propulsion engines:
  - marine engines
  - railway engines
- manufacture of pistons, piston rings, carburettors and such for all internal combustion engines, diesel engines etc.
- manufacture of inlet and exhaust valves of internal combustion engines
- manufacture of turbines and parts thereof:
  - steam turbines and other vapour turbines
  - hydraulic turbines, waterwheels and regulators thereof
  - wind turbines
  - gas turbines, except turbojets or turbo propellers for aircraft propulsion
- manufacture of boiler-turbine sets
- manufacture of turbine-generator sets
- manufacture of engines for industrial application

This class excludes:

- *manufacture of electric generators (except turbine generator sets), see 27.11*
- *manufacture of prime mover generator sets (except turbine generator sets), see 27.11*
- *manufacture of electrical equipment and components of internal combustion engines, see 29.31*
- *manufacture of motor vehicle, aircraft or cycle propulsion engines, see 29.10, 30.30, 30.91*
- *manufacture of turbojets and turbo propellers, see 30.30*

28.12   Manufacture of fluid power equipment

This class includes:

- manufacture of hydraulic and pneumatic components (including hydraulic pumps, hydraulic motors, hydraulic and pneumatic cylinders, hydraulic and pneumatic valves, hydraulic and pneumatic hoses and fittings)
- manufacture of air preparation equipment for use in pneumatic systems
- manufacture of fluid power systems
- manufacture of hydraulic transmission equipment
- manufacture of hydrostatic transmissions

This class excludes:

- *manufacture of compressors, see 28.13*
- *manufacture of pumps for non-hydraulic applications, see 28.13*
- *manufacture of valves for non-fluid power applications, see 28.14*
- *manufacture of mechanical transmission equipment, see 28.15*

28.13   Manufacture of other pumps and compressors

28.13/1   Manufacture of pumps

This subclass includes:

- manufacture of air or vacuum pumps
- manufacture of pumps for liquids whether or not fitted with a measuring device
- manufacture of pumps designed for fitting to internal combustion engines: oil, water and fuel pumps for motor vehicles etc.

This subclass also includes:

- manufacture of hand pumps

This subclass excludes:

- *manufacture of hydraulic and pneumatic equipment, see 28.12*

28.13/2   Manufacture of compressors

This subclass includes:

- manufacture of air or other gas compressors

This subclass excludes:

- *manufacture of hydraulic and pneumatic equipment, see 28.12*

28.14   Manufacture of other taps and valves

This class includes:

- manufacture of industrial taps and valves, including regulating valves and intake taps
- manufacture of sanitary taps and valves
- manufacture of heating taps and valves

This class excludes:

- *manufacture of valves of unhardened vulcanised rubber, glass or of ceramic materials, see 22.19, 23.19 or 23.44*
- *manufacture of inlet and exhaust valves of internal combustion engines, see 28.11*
- *manufacture of hydraulic and pneumatic valves and air preparation equipment for use in pneumatic systems, see 28.12*

28.15   Manufacture of bearings, gears, gearing and driving elements

This class includes:

- manufacture of ball and roller bearings and parts thereof
- manufacture of mechanical power transmission equipment:
  - transmission shafts and cranks: camshafts, crankshafts, cranks etc.
  - bearing housings and plain shaft bearings

- manufacture of gears, gearing and gear boxes and other speed changers
- manufacture of clutches and shaft couplings
- manufacture of flywheels and pulleys
- manufacture of articulated link chain
- manufacture of power transmission chain

This class excludes:

- *manufacture of other chain, see 25.93*
- *manufacture of hydraulic transmission equipment, see 28.12*
- *manufacture of hydrostatic transmissions, see 28.12*
- *manufacture of (electromagnetic) clutches, see 29.31*
- *manufacture of sub-assemblies of power transmission equipment identifiable as parts of vehicles or aircraft, see divisions 29 and 30*

28.2　　　Manufacture of other general-purpose machinery

28.21　　　Manufacture of ovens, furnaces and furnace burners

This class includes:

- manufacture of electrical and other industrial and laboratory furnaces and ovens, including incinerators
- manufacture of burners
- manufacture of permanently mounted electric space heaters, electric swimming pool heaters
- manufacture of permanently mounted non-electric household heating equipment, such as solar heating, steam heating, oil heat and similar furnaces and heating equipment
- manufacture of electric household-type furnaces (electric forced air furnaces, heat pumps, etc.), non-electric household forced air furnaces

This class also includes:

- manufacture of mechanical stokers, grates, ash dischargers etc.

This class excludes:

- *manufacture of household ovens, see 27.51*
- *manufacture of agricultural dryers, see 28.93*
- *manufacture of bakery ovens, see 28.93*
- *manufacture of dryers for wood, paper pulp, paper or paperboard, see 28.99*
- *manufacture of medical, surgical or laboratory sterilisers, see 32.50*
- *manufacture of (dental) laboratory furnaces, see 32.50*

28.22　　　Manufacture of lifting and handling equipment

This class includes:

- manufacture of hand-operated or power-driven lifting, handling, loading or unloading machinery:
  - pulley tackle and hoists, winches, capstans and jacks
  - derricks, cranes, mobile lifting frames, straddle carriers etc.
  - works trucks, whether or not fitted with lifting or handling equipment, whether or not self-propelled, of the type used in factories (including hand trucks and wheelbarrows)
  - mechanical manipulators and industrial robots specifically designed for lifting, handling, loading or unloading
- manufacture of conveyors, teleferics etc.
- manufacture of lifts, escalators and moving walkways
- manufacture of parts specialised for lifting and handling equipment

This class excludes:

- *manufacture of industrial robots for multiple uses, see 28.99*
- *manufacture of continuous-action elevators and conveyors for underground use, see 28.92*
- *manufacture of mechanical shovels, excavators and shovel loaders, see 28.92*
- *manufacture of floating cranes, railway cranes, crane-lorries, see 30.11, 30.20*
- *installation of lifts and elevators, see 43.29*

28.23　　　Manufacture of office machinery and equipment (except computers and peripheral equipment)

This class includes:

- manufacture of calculating machines
- manufacture of adding machines, cash registers
- manufacture of calculators, electronic or not

- manufacture of postage meters, mail handling machines (envelope stuffing, sealing and addressing machinery; opening, sorting, scanning), collating machinery
- manufacture of typewriters
- manufacture of stenography machines
- manufacture of office-type binding equipment (i.e. plastic or tape binding)
- manufacture of cheque writing machines
- manufacture of coin counting and coin wrapping machinery
- manufacture of pencil sharpeners
- manufacture of staplers and staple removers
- manufacture of voting machines
- manufacture of tape dispensers
- manufacture of hole punches
- manufacture of cash registers, mechanically operated
- manufacture of photocopy machines
- manufacture of toner cartridges
- manufacture of blackboards; white boards and marker boards
- manufacture of dictating machines

This class excludes:

- *manufacture of computers and peripheral equipment, see 26.20*

28.24 Manufacture of power-driven hand tools

This class includes:

- manufacture of hand tools, with self-contained electric or non-electric motor or pneumatic drive, such as:
  - circular or reciprocating saws
  - chain saws
  - drills and hammer drills
  - hand held power sanders
  - pneumatic nailers
  - buffers
  - routers
  - grinders
  - staplers
  - pneumatic rivet guns
  - planers
  - shears and nibblers
  - impact wrenches
  - powder actuated nailers

This class excludes:

- *manufacture of interchangeable tools for hand tools, see 25.73*
- *manufacture of electrical hand held soldering and welding equipment, see 27.90*

28.25 Manufacture of non-domestic cooling and ventilation equipment

This class includes:

- manufacture of refrigerating or freezing industrial equipment, including assemblies of components
- manufacture of air-conditioning machines, including for motor vehicles
- manufacture of non-domestic fans
- manufacture of heat exchangers
- manufacture of machinery for liquefying air or gas
- manufacture of attic ventilation fans (gable fans, roof ventilators, etc.)

This class excludes:

- *manufacture of domestic refrigerating or freezing equipment, see 27.51*
- *manufacture of domestic fans, see 27.51*

28.29 Manufacture of other general-purpose machinery n.e.c.

This class includes:

- manufacture of weighing machinery (other than sensitive laboratory balances):
  - household and shop scales, platform scales, scales for continuous weighing, weighbridges, weights etc.

C

– manufacture of filtering or purifying machinery and apparatus for liquids
– manufacture of equipment for projecting, dispersing or spraying liquids or powders:
  ■ spray guns, fire extinguishers, sandblasting machines, steam cleaning machines etc.
– manufacture of packing and wrapping machinery:
  ■ filling, closing, sealing, capsuling or labelling machines etc.
– manufacture of machinery for cleaning or drying bottles and for aerating beverages
– manufacture of distilling or rectifying plant for petroleum refineries, chemical industries, beverage industries etc.
– manufacture of gas generators
– manufacture of calendaring or other rolling machines and cylinders thereof (except for metal and glass)
– manufacture of centrifuges (except cream separators and clothes dryers)
– manufacture of gaskets and similar joints made of a combination of materials or layers of the same material
– manufacture of automatic goods vending machines
– manufacture of levels, tape measures and similar hand tools, machinists' precision tools (except optical)
– manufacture of non-electrical welding and soldering equipment
– manufacture of cooling towers and similar for direct cooling by means of re-circulated water

This class excludes:

– *manufacture of sensitive (laboratory-type) balances, see 26.51*
– *manufacture of domestic refrigerating or freezing equipment, see 27.51*
– *manufacture of domestic fans, see 27.51*
– *manufacture of electrical welding and soldering equipment, see 27.90*
– *manufacture of agricultural spraying machinery, see 28.30*
– *manufacture of metal or glass rolling machinery and cylinders thereof, see 28.91, 28.99*
– *manufacture of agricultural dryers, see 28.93*
– *manufacture of machinery for filtering or purifying food, see 28.93*
– *manufacture of cream separators, see 28.93*
– *manufacture of commercial clothes dryers, see 28.94*
– *manufacture of textile printing machinery, see 28.94*

28.3　　　　　Manufacture of agricultural and forestry machinery

28.30　　　　　Manufacture of agricultural and forestry machinery

28.30/1　　　　Manufacture of agricultural tractors

This subclass includes:

– manufacture of tractors used in agriculture and forestry
– manufacture of walking (pedestrian-controlled) tractors

This subclass excludes:

– *manufacture of road tractors for semi-trailers, see 29.10*

28.30/2　　　　Manufacture of agricultural and forestry machinery (other than agricultural tractors)

This subclass includes:

– manufacture of mowers, including lawnmowers
– manufacture of agricultural self-loading or self-unloading trailers or semi-trailers
– manufacture of agricultural machinery for soil preparation, planting or fertilizing:
  ■ ploughs, manure spreaders, seeders, harrows etc.
– manufacture of harvesting or threshing machinery:
  ■ harvesters, threshers, sorters etc.
– manufacture of milking machines
– manufacture of spraying machinery for agricultural use
– manufacture of diverse agricultural machinery:
  ■ poultry-keeping machinery, bee-keeping machinery, equipment for preparing fodder etc.
  ■ machines for cleaning, sorting or grading eggs, fruit etc.

This subclass excludes:

– *manufacture of non-power-driven agricultural hand tools, see 25.73*
– *manufacture of conveyors for farm use, see 28.22*
– *manufacture of power-driven hand tools, see 28.24*
– *manufacture of cream separators, see 28.93*

– *manufacture of machinery to clean, sort or grade seed, grain or dried leguminous vegetables, see 28.93*

– *manufacture of road trailers or semi-trailers, see 29.20*

28.4   Manufacture of metal forming machinery and machine tools

This group includes the manufacture of metal forming machinery and machine tools, e.g. manufacture of machine tools for working metals and other materials (wood, bone, stone, hard rubber, hard plastics, cold glass etc.), including those using a laser beam, ultrasonic waves, plasma arc, magnetic pulse etc.

28.41   Manufacture of metal forming machinery

This class includes:

– manufacture of machine tools for working metals, including those using a laser beam, ultrasonic waves, plasma arc, magnetic pulse etc.
– manufacture of machine tools for turning, drilling, milling, shaping, planing, boring, grinding etc.
– manufacture of stamping or pressing machine tools
– manufacture of punch presses, hydraulic presses, hydraulic brakes, drop hammers, forging machines etc.
– manufacture of draw-benches, thread rollers or machines for working wires

This class excludes:

– *manufacture of interchangeable tools, see 25.73*
– *manufacture of electrical welding and soldering machines, see 27.90*

28.49   Manufacture of other machine tools

This class includes:

– manufacture of machine tools for working wood, bone, stone, hard rubber, hard plastics, cold glass etc., including those using a laser beam, ultrasonic waves, plasma arc, magnetic pulse etc.
– manufacture of work holders for machine tools
– manufacture of dividing heads and other special attachments for machine tools
– manufacture of stationary machines for nailing, stapling, glueing or otherwise assembling wood, cork, bone, hard rubber or plastics etc.
– manufacture of stationary rotary or rotary percussion drills, filing machines, riveters, sheet metal cutters etc.
– manufacture of presses for the manufacture of particle board and the like
– manufacture of electroplating machinery

This class also includes:

– manufacture of parts and accessories for the machine tools listed

This class excludes:

– *manufacture of interchangeable tools for machine tools (drills, punches, dies, taps, milling cutters, turning tools, saw blades, cutting knives etc.), see 25.73*
– *manufacture of electric hand held soldering irons and soldering guns, see 27.90*
– *manufacture of power-driven hand tools, see 28.24*
– *manufacture of machinery used in metal mills or foundries, see 28.91*
– *manufacture of machinery for mining and quarrying, see 28.92*

28.9   Manufacture of other special-purpose machinery

This group includes the manufacture of special-purpose machinery, i.e. machinery for exclusive use in a specific industry or a small cluster of specific industries. While most of these are used in other manufacturing processes, such as food manufacturing or textile manufacturing, this group also includes the manufacture of machinery specific for other (non-manufacturing industries), such as aircraft launching gear or amusement park equipment.

28.91   Manufacture of machinery for metallurgy

This class includes:

– manufacture of machines and equipment for handling hot metals:
  ■ converters, ingot moulds, ladles, casting machines
– manufacture of metal-rolling mills and rolls for such mills

This class excludes:

– *manufacture of draw-benches, see 28.41*
– *manufacture of moulding boxes and moulds (except ingot moulds), see 25.73*
– *manufacture of machines for forming foundry moulds, see 28.99*

28.92   Manufacture of machinery for mining, quarrying and construction

28.92/1   Manufacture of machinery for mining

This subclass includes:

– manufacture of continuous-action elevators and conveyors for underground use
– manufacture of boring, cutting, sinking and tunnelling machinery (whether or not for underground use)
– manufacture of machinery for treating minerals by screening, sorting, separating, washing, crushing etc.
– manufacture of track laying tractors and tractors used in mining

This subclass excludes:

– *manufacture of lifting and handling equipment, see 28.22*
– *manufacture of other tractors, see 28.30, 29.10*
– *manufacture of machine tools for working stone, including machines for splitting or clearing stone, see 28.49*
– *manufacture of concrete-mixer lorries, see 29.10*

28.92/2   Manufacture of earthmoving equipment

This subclass includes:

– manufacture of earth-moving machinery:
  ■ bulldozers, angle-dozers, graders, scrapers, levellers, mechanical shovels, shovel loaders etc.
– manufacture of bulldozer and angle-dozer blades
– manufacture of off-road dumping trucks

28.92/3   Manufacture of equipment for concrete crushing and screening roadworks

This subclass includes:

– manufacture of concrete and mortar mixers
– manufacture of pile-drivers and pile-extractors, mortar spreaders, bitumen spreaders, concrete surfacing machinery etc.
– manufacture of track laying tractors and tractors used in construction

28.93   Manufacture of machinery for food, beverage and tobacco processing

This class includes:

– manufacture of agricultural dryers
– manufacture of machinery for the dairy industry:
  ■ cream separators
  ■ milk processing machinery (e.g. homogenisers)
  ■ milk converting machinery (e.g. butter churns, butter workers and moulding machines)
  ■ cheese-making machines (e.g. homogenisers, moulders, presses) etc.
– manufacture of machinery for the grain milling industry:
  ■ machinery to clean, sort or grade seeds, grain or dried leguminous vegetables (winnowers, sieving belts, separators, grain brushing machines etc.)
  ■ machinery to produce flour and meal etc. (grinding mills, feeders, sifters, bran cleaners, blenders, rice hullers, pea splitters)
– manufacture of presses, crushers etc. used to make wine, cider, fruit juices etc.
– manufacture of machinery for the bakery industry or for making macaroni, spaghetti or similar products:
  ■ bakery ovens, dough mixers, dough-dividers, moulders, slicers, cake depositing machines etc.
– manufacture of machines and equipment to process diverse foods:
  ■ machinery to make confectionery, cocoa or chocolate; to manufacture sugar; for breweries; to process meat or poultry, to prepare fruit, nuts or vegetables; to prepare fish, shellfish or other seafood
  ■ machinery for filtering and purifying
  ■ other machinery for the industrial preparation or manufacture of food or drink
– manufacture of machinery for the extraction or preparation of animal or vegetable fats or oils
– manufacture of machinery for the preparation of tobacco and for the making of cigarettes or cigars, or for pipe or chewing tobacco or snuff
– manufacture of machinery for the preparation of food in hotels and restaurants

This class excludes:

– *manufacture of food and milk irradiation equipment, see 26.60*
– *manufacture of packing, wrapping and weighing machinery, see 28.29*
– *manufacture of cleaning, sorting or grading machinery for eggs, fruit or other crops (except seeds, grains and dried leguminous vegetables), see 28.30*

C

28.94   Manufacture of machinery for textile, apparel and leather production

This class includes:

– manufacture of textile machinery:
  - machines for preparing, producing, extruding, drawing, texturing or cutting man-made textile fibres, materials or yarns
  - machines for preparing textile fibres: cotton gins, bale breakers, garnetters, cotton spreaders, wool scourers, wool carbonisers, combs, carders, roving frames etc.
  - spinning machines
  - machines for preparing textile yarns: reelers, warpers and related machines
  - weaving machines (looms), including hand looms
  - knitting machines
  - machines for making knotted net, tulle, lace, braid etc.
– manufacture of auxiliary machines or equipment for textile machinery:
  - dobbies, jacquards, automatic stop motions, shuttle changing mechanisms, spindles and spindle flyers etc.
– manufacture of textile printing machinery
– manufacture of machinery for fabric processing:
  - machinery for washing, bleaching, dyeing, dressing, finishing, coating or impregnating textile fabrics
  - manufacture of machines for reeling, unreeling, folding, cutting or pinking textile fabrics
– manufacture of laundry machinery:
  - ironing machines, including fusing presses
  - commercial washing and drying machines
  - dry-cleaning machines
– manufacture of sewing machines, sewing machine heads and sewing machine needles (whether or not for household use)
– manufacture of machines for producing or finishing felt or non-wovens
– manufacture of leather machines:
  - machinery for preparing, tanning or working hides, skins or leather
  - machinery for making or repairing footwear or other articles of hides, skins, leather or fur skins

This class excludes:

– *manufacture of paper or paperboard cards for use on jacquard machines, see 17.29*
– *manufacture of domestic washing and drying machines, see 27.51*
– *manufacture of calendaring machines, see 28.29*
– *manufacture of machines used in bookbinding, see 28.99*

28.95   Manufacture of machinery for paper and paperboard production

This class includes:

– manufacture of machinery for making paper pulp
– manufacture of paper and paperboard making machinery
– manufacture of machinery producing articles of paper or paperboard

28.96   Manufacture of plastics and rubber machinery

This class includes:

– manufacture of machinery for working soft rubber or plastics or for the manufacture of products of these materials:
  - extruders, moulders, pneumatic tyre making or retreading machines and other machines for making a specific rubber or plastic product

28.99   Manufacture of other special-purpose machinery n.e.c.

This class includes the manufacture of special-purpose machinery not elsewhere classified.

This class includes:

– manufacture of dryers for wood, paper pulp, paper or paperboard and other materials (except for agricultural products and textiles)
– manufacture of printing and bookbinding machines and machines for activities supporting printing on a variety of materials
– manufacture of machinery for producing tiles, bricks, shaped ceramic pastes, pipes, graphite electrodes, blackboard chalk etc.
– manufacture of semi-conductor manufacturing machinery
– manufacture of industrial robots performing multiple tasks for special purposes
– manufacture of diverse special purpose machinery and equipment:
  - machines to assemble electric or electronic lamps, tubes (valves) or bulbs
  - machines for production or hot-working of glass or glassware, glass fibre or yarn

　　　　　　■ machinery or apparatus for isotopic separation
　　　　– manufacture of tyre alignment and balancing equipment; balancing equipment (except wheel balancing)
　　　　– manufacture of central greasing systems
　　　　– manufacture of aircraft launching gear, aircraft carrier catapults and related equipment
　　　　– manufacture of tanning beds
　　　　– manufacture of automatic bowling alley equipment (e.g. pin-setters)
　　　　– manufacture of roundabouts, swings, shooting galleries and other fairground amusements

This class excludes:

– *manufacture of wheel balancing equipment, see 26.51*

– *manufacture of household appliances, see 27.5*

– *manufacture of photocopy machines etc., see 28.23*

– *manufacture of machinery or equipment to work hard rubber, hard plastics or cold glass, see 28.49*

– *manufacture of ingot moulds, see 28.91*

29　Manufacture of motor vehicles, trailers and semi-trailers

This division includes the manufacture of motor vehicles for transporting passengers or freight. The manufacture of various parts and accessories, as well as the manufacture of trailers and semi-trailers, is included here.

The maintenance and repair of motor vehicles produced in this division are classified in 45.20.

29.1　Manufacture of motor vehicles

29.10　Manufacture of motor vehicles

This class includes:

– manufacture of passenger cars
– manufacture of commercial vehicles:
　　■ vans, lorries, on-road tractor units for semi-trailers etc.
– manufacture of buses, trolley-buses and coaches
– manufacture of motor vehicle engines
– manufacture of chassis for motor vehicles
– manufacture of other motor vehicles:
　　■ snowmobiles, golf carts, amphibious vehicles
　　■ fire engines, street sweepers, travelling libraries, armoured cars etc.
　　■ concrete-mixer lorries
– ATVs, go-carts and similar including race cars

This class also includes:

– factory rebuilding of motor vehicle engines

This class excludes:

– *manufacture of electric motors (except starting motors), see 27.11*

– *manufacture of lighting equipment for motor vehicles, see 27.40*

– *manufacture of pistons, piston rings and carburettors, see 28.11*

– *manufacture of agricultural tractors, see 28.30*

– *manufacture of tractors used in construction or mining, see 28.92*

– *manufacture of off-road dumping trucks, see 28.92*

– *manufacture of bodies for motor vehicles, see 29.20*

– *manufacture of electrical parts for motor vehicles, see 29.31*

– *manufacture of parts and accessories for motor vehicles, see 29.32*

– *manufacture of tanks and other military fighting vehicles, see 30.40*

– *maintenance, and repair of motor vehicles, see 45.20*

29.2　Manufacture of bodies (coachwork) for motor vehicles; manufacture of trailers and semi-trailers

29.20　Manufacture of bodies (coachwork) for motor vehicles; manufacture of trailers and semi-trailers

29.20/1　Manufacture of bodies (coachwork) for motor vehicles (except caravans)

This subclass includes:

– manufacture of bodies, including cabs for motor vehicles
– outfitting of all types of motor vehicles (except caravans)

This subclass excludes:

– *manufacture of parts and accessories of bodies for motor vehicles, see 29.32*
– *manufacture of vehicles drawn by animals, see 30.99*

C

29.20/2   Manufacture of trailers and semi-trailers

This subclass includes:

– manufacture of trailers and semi-trailers:
  ■ tankers, removal trailers etc.
– manufacture of containers for carriage by one or more modes of transport

This subclass excludes:

– *manufacture of trailers and semi-trailers specially designed for use in agriculture, see 28.30*
– *manufacture of parts and accessories of bodies for motor vehicles, see 29.32*
– *manufacture of vehicles drawn by animals, see 30.99*

29.20/3   Manufacture of caravans

This subclass includes:

– manufacture of caravan trailers and the like
– outfitting of caravans

This subclass excludes:

– *manufacture of parts and accessories of bodies for motor vehicles, see 29.32*
– *manufacture of vehicles drawn by animals, see 30.99*

29.3   **Manufacture of parts and accessories for motor vehicles**

29.31   Manufacture of electrical and electronic equipment for motor vehicles

This class includes:

– manufacture of motor vehicle electrical equipment, such as generators, alternators, spark plugs, ignition wiring harnesses, power window and door systems, assembly of purchased gauges into instrument panels, voltage regulators, etc.

This class excludes:

– *manufacture of batteries for vehicles, see 27.20*
– *manufacture of lighting equipment for motor vehicles, see 27.40*
– *manufacture of pumps for motor vehicles and engines, see 28.13*

29.32   Manufacture of other parts and accessories for motor vehicles

This class includes:

– manufacture of diverse parts and accessories for motor vehicles:
  ■ brakes, gearboxes, axles, road wheels, suspension shock absorbers, radiators, silencers, exhaust pipes, catalytic converters, clutches, steering wheels, steering columns and steering boxes
– manufacture of parts and accessories of bodies for motor vehicles:
  ■ safety belts, airbags, doors, bumpers
– manufacture of car seats

This class excludes:

– *manufacture of tyres, see 22.11*
– *manufacture of rubber hoses and belts and other rubber products, see 22.19*
– *manufacture of pistons, piston rings and carburettors, see 28.11*
– *maintenance, repair and alteration of motor vehicles, see 45.20*

30   **Manufacture of other transport equipment**

This division includes the manufacture of transportation equipment such as ship building and boat manufacturing, the manufacture of railway rolling stock and locomotives, air and spacecraft and the manufacture of parts thereof.

30.1   **Building of ships and boats**

This group includes the building of ships, boats and other floating structures for transportation and other commercial purposes, as well as for sports and recreational purposes.

30.11   Building of ships and floating structures

This class includes the building of ships, except vessels for sports or recreation, and the construction of floating structures:

This class includes:

- building of commercial vessels:
  - passenger vessels, ferry-boats, cargo ships, tankers, tugs etc.
- building of warships
- building of fishing boats and fish-processing factory vessels

This class also includes:

- building of hovercraft (except recreation-type hovercraft)
- construction of drilling platforms, floating or submersible
- construction of floating structures:
  - floating docks, pontoons, coffer-dams, floating landing stages, buoys, floating tanks, barges, lighters, floating cranes, non-recreational inflatable rafts etc.
- manufacture of sections for ships and floating structures

This class excludes:

- *manufacture of parts of vessels, other than major hull assemblies:*
  - *manufacture of sails, see 13.92*
  - *manufacture of ships' propellers, see 25.99*
  - *manufacture of iron or steel anchors, see 25.99*
  - *manufacture of marine engines, see 28.11*
- *manufacture of navigational instruments, see 26.51*
- *manufacture of lighting equipment for ships, see 27.40*
- *manufacture of amphibious motor vehicles, see 29.10*
- *manufacture of inflatable boats or rafts for recreation, see 30.12*
- *specialised repair and maintenance of ships and floating structures, see 33.15*
- *ship-breaking, see 38.31*
- *interior installation of boats, see 43.3*

30.12 Building of pleasure and sporting boats

This class includes:

- manufacture of inflatable boats and rafts
- building of sailboats with or without auxiliary motor
- building of motor boats
- building of recreation-type hovercraft
- manufacture of personal watercraft
- manufacture of other pleasure and sporting boats:
  - canoes, kayaks, rowing boats, skiffs

This class excludes:

- *manufacture of parts of pleasure and sporting boats:*
  - *manufacture of sails, see 13.92*
  - *manufacture of iron or steel anchors, see 25.99*
  - *manufacture of marine engines, see 28.11*
- *manufacture of sailboards and surfboards, see 32.30*
- *maintenance and repair of pleasure boats, see 33.15*

30.2 Manufacture of railway locomotives and rolling stock

30.20 Manufacture of railway locomotives and rolling stock

This class includes:

- manufacture of electric, diesel, steam and other rail locomotives
- manufacture of self-propelled railway or tramway coaches, vans and trucks, maintenance or service vehicles
- manufacture of railway or tramway rolling stock, not self-propelled:
  - passenger coaches, goods vans, tank wagons, self-discharging vans and wagons, workshop vans, crane vans, tenders etc.
- manufacture of specialised parts of railway or tramway locomotives or of rolling stock:
  - bogies, axles and wheels, brakes and parts of brakes; hooks and coupling devices, buffers and buffer parts; shock absorbers; wagon and locomotive frames; bodies; corridor connections etc.

This class also includes:

- manufacture of mining locomotives and mining rail cars

C

  – manufacture of mechanical and electromechanical signalling, safety and traffic control equipment for railways, tramways, inland waterways, roads, parking facilities, airfields etc.
  – manufacture of railway car seats

This class excludes:

  – *manufacture of unassembled rails, see 24.10*
  – *manufacture of assembled railway track fixtures, see 25.99*
  – *manufacture of electric motors, see 27.11*
  – *manufacture of electrical signalling, safety or traffic-control equipment, see 27.90*
  – *manufacture of engines and turbines, see 28.11*

30.3   **Manufacture of air and spacecraft and related machinery**

30.30   **Manufacture of air and spacecraft and related machinery**

This class includes:

  – manufacture of aeroplanes for the transport of goods or passengers, for use by the defence forces, for sport or other purposes
  – manufacture of helicopters
  – manufacture of gliders, hang-gliders
  – manufacture of dirigibles and hot air balloons
  – manufacture of parts and accessories of the aircraft of this class:
    ■ major assemblies such as fuselages, wings, doors, control surfaces, landing gear, fuel tanks, nacelles etc.
    ■ airscrews, helicopter rotors and propelled rotor blades
    ■ motors and engines of a kind typically found on aircraft
    ■ parts of turbojets and turboprops for aircraft
  – manufacture of ground flying trainers
  – manufacture of spacecraft and launch vehicles, satellites, planetary probes, orbital stations, shuttles
  – manufacture of intercontinental ballistic missiles (ICBM)

This class also includes:

  – overhaul and conversion of aircraft or aircraft engines
  – manufacture of aircraft seats

This class excludes:

  – *manufacture of parachutes, see 13.92*
  – *manufacture of military ordnance and ammunition, see 25.40*
  – *manufacture of telecommunication equipment for satellites, see 26.30*
  – *manufacture of aircraft instrumentation and aeronautical instruments, see 26.51*
  – *manufacture of air navigation systems, see 26.51*
  – *manufacture of lighting equipment for aircraft, see 27.40*
  – *manufacture of ignition parts and other electrical parts for internal combustion engines, see 27.90*
  – *manufacture of pistons, piston rings and carburettors, see 28.11*
  – *manufacture of aircraft launching gear, aircraft carrier catapults and related equipment, see 28.99*

30.4   **Manufacture of military fighting vehicles**

30.40   **Manufacture of military fighting vehicles**

This class includes:

  – manufacture of tanks
  – manufacture of armoured amphibious military vehicles
  – manufacture of other military fighting vehicles

This class excludes:

  – *manufacture of weapons and ammunitions, see 25.40*

30.9   **Manufacture of transport equipment n.e.c.**

This group includes the manufacture of transport equipment other than motor vehicles and rail, water, air or space transport equipment and military vehicles.

30.91   **Manufacture of motorcycles**

This class includes:

  – manufacture of motorcycles, mopeds and cycles fitted with an auxiliary engine

- manufacture of engines for motorcycles
- manufacture of sidecars
- manufacture of parts and accessories for motorcycles

This class excludes:

- *manufacture of bicycles, see 30.92*
- *manufacture of invalid carriages, see 30.92*

30.92   Manufacture of bicycles and invalid carriages

This class includes:

- manufacture of non-motorised bicycles and other cycles, including (delivery) tricycles, tandems, children's bicycles and tricycles
- manufacture of parts and accessories of bicycles
- manufacture of invalid carriages/wheelchairs with or without motor
- manufacture of parts and accessories of invalid carriages
- manufacture of baby carriages

This class excludes:

- *manufacture of bicycles with auxiliary motor, see 30.91*
- *manufacture of wheeled toys designed to be ridden, including plastic bicycles and tricycles, see 32.40*

30.99   Manufacture of other transport equipment n.e.c.

This class includes:

- manufacture of hand-propelled vehicles: luggage trucks, handcarts, sledges, shopping carts etc.
- manufacture of vehicles drawn by animals: sulkies, donkey-carts, hearses etc.

This class excludes:

- *works trucks, whether or not fitted with lifting or handling equipment, whether or not self-propelled, of the type used in factories (including hand trucks and wheelbarrows), see 28.22*
- *decorative restaurant carts, such as a dessert cart, food wagons, see 31.01*

31   **Manufacture of furniture**

This division includes the manufacture of furniture and related products of any material except stone, concrete and ceramic. The processes used in the manufacture of furniture are standard methods of forming materials and assembling components, including cutting, moulding and laminating. The design of the article, for both aesthetic and functional qualities, is an important aspect of the production process.

Some of the processes used in furniture manufacturing are similar to processes that are used in other segments of manufacturing. For example, cutting and assembly occurs in the production of wood trusses that are classified in division 16 (Manufacture of wood and wood products). However, the multiple processes distinguish wood furniture manufacturing from wood product manufacturing. Similarly, metal furniture manufacturing uses techniques that are also employed in the manufacturing of roll-formed products classified in division 25 (Manufacture of fabricated metal products). The moulding process for plastics furniture is similar to the moulding of other plastics products. However, the manufacture of plastics furniture tends to be a specialised activity.

31.0   **Manufacture of furniture**

31.01   Manufacture of office and shop furniture

This class includes the manufacture of furniture of any kind (other than domestic), of any material (except stone, concrete or ceramic) for any place and for various purposes.

This class includes:

- manufacture of chairs and seats for offices, workrooms, hotels, restaurants and public premises
- manufacture of chairs and seats for theatres, cinemas and the like
- manufacture of special furniture for shops: counters, display cases, shelves etc.
- manufacture of office furniture
- manufacture of laboratory benches, stools, and other laboratory seating, laboratory furniture (e.g. cabinets and tables)
- manufacture of furniture for churches, schools, restaurants

This class also includes:

- decorative restaurant carts, such as a dessert cart, food wagons

This class excludes:

- *blackboards, see 28.23*

C

- *manufacture of car seats, see 29.32*
- *manufacture of railway car seats, see 30.20*
- *manufacture of aircraft seats, see 30.30*
- *manufacture of medical, surgical, dental or veterinary furniture, see 32.50*
- *modular furniture attachment and installation, partition installation, laboratory equipment furniture installation, see 43.32*

31.02 Manufacture of kitchen furniture

This class includes:

- manufacture of kitchen furniture

31.03 Manufacture of mattresses

This class includes:

- manufacture of mattresses:
  - mattresses fitted with springs or stuffed or internally fitted with a supporting material
  - uncovered cellular rubber or plastic mattresses
- manufacture of mattress supports

This class excludes:

- *manufacture of inflatable rubber mattresses, see 22.19*
- *manufacture of rubber waterbed mattresses, see 22.19*

31.09 Manufacture of other furniture

This class includes:

- manufacture of sofas, sofa beds and sofa sets
- manufacture of garden chairs and seats
- manufacture of furniture for bedrooms, living rooms, gardens etc.
- manufacture of cabinets for sewing machines, televisions etc.

This class also includes:

- finishing such as upholstery of chairs and seats
- finishing of furniture such as spraying, painting, French polishing and upholstering

This class excludes:

- *manufacture of pillows, pouffes, cushions, quilts and eiderdowns, see 13.92*
- *manufacture of furniture of ceramics, concrete and stone, see 23.42, 23.69, 23.70*
- *manufacture of lighting fittings or lamps, see 27.40*
- *manufacture of car seats, see 29.32*
- *manufacture of railway car seats, see 30.20*
- *manufacture of aircraft seats, see 30.30*
- *reupholstering and restoring of furniture, see 95.24*

32 Other manufacturing

This division includes the manufacture of a variety of goods not covered in other parts of the classification. Since this is a residual division, production processes, input materials and use of the produced goods can vary widely and usual criteria for grouping classes into divisions have not been applied here.

32.1 Manufacture of jewellery, bijouterie and related articles

This group includes the manufacture of jewellery and imitation jewellery articles.

32.11 Striking of coins

This class includes:

- manufacture of coins, including coins for use as legal tender, whether or not of precious metal

32.12 Manufacture of jewellery and related articles

This class includes:

- production of worked pearls
- production of precious and semi-precious stones in the worked state, including the working of industrial quality stones and synthetic or reconstructed precious or semi-precious stones
- working of diamonds

- manufacture of jewellery of precious metal or of base metals clad with precious metals, or precious or semi-precious stones, or of combinations of precious metal and precious or semi-precious stones or of other materials
- manufacture of goldsmiths' articles of precious metals or of base metals clad with precious metals:
  - dinnerware, flatware, hollowware, toilet articles, office or desk articles, articles for religious use etc.
- manufacture of technical or laboratory articles of precious metal (except instruments and parts thereof): crucibles, spatulas, electroplating anodes etc.
- manufacture of precious metal watch bands, wristbands, watch straps and cigarette cases

This class also includes:

- engraving of personal precious and non-precious metal products

This class excludes:

- *manufacture of non-metal watch bands (fabric, leather, plastic etc.), see 15.12*
- *manufacture of articles of base metal plated with precious metal (except imitation jewellery), see division 25*
- *manufacture of watch cases, see 26.52*
- *manufacture of (non-precious) metal watch bands, see 32.13*
- *manufacture of imitation jewellery, see 32.13*
- *repair of jewellery, see 95.25*

32.13　　Manufacture of imitation jewellery and related articles

This class includes:

- manufacture of costume or imitation jewellery:
  - rings, bracelets, necklaces, and similar articles of jewellery made from base metals plated with precious metals
  - jewellery containing imitation stones such as imitation gems stones, imitation diamonds, and similar
- manufacture of metal watch bands (except precious metal)

This class excludes:

- *manufacture of jewellery made from precious metals or clad with precious metals, see 32.12*
- *manufacture of jewellery containing genuine gem stones, see 32.12*
- *manufacture of precious metal watch bands, see 32.12*

32.2　　Manufacture of musical instruments

32.20　　Manufacture of musical instruments

This class includes:

- manufacture of stringed instruments
- manufacture of keyboard stringed instruments, including automatic pianos
- manufacture of keyboard pipe organs, including harmoniums and similar keyboard instruments with free metal reeds
- manufacture of accordions and similar instruments, including mouth organs
- manufacture of wind instruments
- manufacture of percussion musical instruments
- manufacture of musical instruments, the sound of which is produced electronically
- manufacture of musical boxes, fairground organs, calliopes etc.
- manufacture of instrument parts and accessories:
  - metronomes, tuning forks, pitch pipes, cards, discs and rolls for automatic mechanical instruments etc.

This class also includes:

- manufacture of whistles, call horns and other mouth-blown sound signalling instruments

This class excludes:

- *reproduction of pre-recorded sound and video tapes and discs, see 18.2*
- *manufacture of microphones, amplifiers, loudspeakers, headphones and similar components, see 26.40*
- *manufacture of record players, tape recorders and the like, see 26.40*
- *manufacture of toy musical instruments, see 32.40*
- *restoring of organs and other historic musical instruments, see 33.19*
- *publishing of pre-recorded sound and video tapes and discs, see 59.20*
- *piano tuning, see 95.29*

32.3　　Manufacture of sports goods

32.30　　Manufacture of sports goods

This class includes the manufacture of sporting and athletic goods (except apparel and footwear).

This class includes:

– manufacture of articles and equipment for sports, outdoor and indoor games, of any material:
  - hard, soft and inflatable balls
  - rackets, bats and clubs
  - skis, bindings and poles
  - ski-boots
  - sailboards and surfboards
  - requisites for sport fishing, including landing nets
  - requisites for hunting, mountain climbing etc.
  - leather sports gloves and sports headgear
  - basins for swimming and padding pools etc.
  - ice skates, roller skates etc.
  - bows and crossbows
  - gymnasium, fitness centre or athletic equipment

This class excludes:

– *manufacture of boat sails, see 13.92*
– *manufacture of sports apparel, see 14.19*
– *manufacture of saddlery and harness, see 15.12*
– *manufacture of whips and riding crops, see 15.12*
– *manufacture of sports footwear, see 15.20*
– *manufacture of sporting weapons and ammunition, see 25.40*
– *manufacture of metal weights as used for weightlifting, see 25.99*
– *manufacture of sports vehicles other than toboggans and the like, see divisions 29 and 30*
– *manufacture of boats, see 30.12*
– *manufacture of billiard tables, see 32.40*
– *manufacture of ear and noise plugs (e.g. for swimming and noise protection), see 32.99*
– *repair of sporting goods, see 95.29*

32.4   **Manufacture of games and toys**

32.40   Manufacture of games and toys

32.40/1   Manufacture of professional and arcade games and toys

This subclass includes:

– manufacture of coin-operated games, billiards, special tables for casino games, etc.
– manufacture of articles for funfair games

This subclass excludes:

– *manufacture of automatic bowling alley equipment (e.g. pin setters), see 28.99*

32.40/9   Manufacture of games and toys (other than professional and arcade games and toys) n.e.c.

This subclass includes the manufacture of dolls, toys and games (including electronic games), scale models and children's vehicles (except metal bicycles and tricycles).

This subclass includes:

– manufacture of dolls and doll garments, parts and accessories
– manufacture of action figures
– manufacture of toy animals
– manufacture of toy musical instruments
– manufacture of playing cards
– manufacture of board games and similar games
– manufacture of electronic games: chess etc.
– manufacture of reduced-size ("scale") models and similar recreational models, electrical trains, construction sets etc.
– manufacture of articles for table or parlour games
– manufacture of wheeled toys designed to be ridden, including plastic bicycles and tricycles
– manufacture of puzzles and similar articles

This subclass excludes:

– *manufacture of video game consoles, see 26.40*
– *manufacture of bicycles, see 30.92*

- *manufacture of articles for jokes and novelties, see 32.99*
- *writing and publishing of software for video game consoles, see 58.21, 62.01*

32.5   **Manufacture of medical and dental instruments and supplies**

32.50   Manufacture of medical and dental instruments and supplies

This class includes the manufacture of laboratory apparatus, surgical and medical instruments, surgical appliances and supplies, dental equipment and supplies, orthodontic goods, dentures, and orthodontic appliances. Included is the manufacture of medical, dental and similar furniture, where the additional specific functions determine the purpose of the product, such as dentist's chairs with built-in hydraulic functions.

This class includes:

- manufacture of surgical drapes and sterile string and tissue
- manufacture of dental fillings and cements (except denture adhesives), dental wax and other dental plaster preparations
- manufacture of bone reconstruction cements
- manufacture of dental laboratory furnaces
- manufacture of laboratory ultrasonic cleaning machinery
- manufacture of laboratory sterilisers
- manufacture of laboratory type distilling apparatus, laboratory centrifuges
- manufacture of medical, surgical, dental or veterinary furniture, such as:
  - operating tables
  - examination tables
  - hospital beds with mechanical fittings
  - dentists' chairs
- manufacture of bone plates and screws, syringes, needles, catheters, cannulae, etc.
- manufacture of dental instruments (including dentists' chairs incorporating dental equipment)
- manufacture of artificial teeth, bridges, etc., made in dental labs
- manufacture of orthopedic and prosthetic devices
- manufacture of glass eyes
- manufacture of medical thermometers
- manufacture of ophthalmic goods, eyeglasses, sunglasses, lenses ground to prescription, contact lenses, safety goggles

This class excludes:

- *manufacture of denture adhesives, see 20.42*
- *manufacture of medical impregnated wadding, dressings etc., see 21.20*
- *manufacture of electromedical and electrotherapeutic equipment, see 26.60*
- *manufacture of wheelchairs, see 30.92*
- *activities of opticians, see 47.78*

32.9   **Manufacturing n.e.c**

32.91   Manufacture of brooms and brushes

This class includes:

- manufacture of brooms and brushes, including brushes constituting parts of machines, hand-operated mechanical floor sweepers, mops and feather dusters, paint brushes, paint pads and rollers, squeegees and other brushes, brooms, mops etc.
- manufacture of shoe and clothes brushes

This class excludes:

- *rubber brushes, see 22.19*

32.99   Other manufacturing n.e.c.

This class includes:

- manufacture of protective safety equipment
  - manufacture of fire-resistant and protective safety clothing
  - manufacture of linemen's safety belts and other belts for occupational use
  - manufacture of cork life preservers
  - manufacture of plastics hard hats and other personal safety equipment of plastics
  - manufacture of fire-fighting protection suits
  - manufacture of metal safety headgear and other metal personal safety devices
  - manufacture of ear and noise plugs (e.g. for swimming and noise protection)
  - manufacture of gas masks

– manufacture of pens and pencils of all kinds whether or not mechanical
– manufacture of pencil leads
– manufacture of date, sealing or numbering stamps, hand-operated devices for printing, or embossing labels, hand printing sets, prepared typewriter ribbons and inked pads
– manufacture of globes
– manufacture of umbrellas, sun-umbrellas, walking sticks, seat-sticks
– manufacture of buttons, press-fasteners, snap-fasteners, press-studs, slide fasteners
– manufacture of cigarette lighters
– manufacture of articles of personal use: smoking pipes, combs, hair slides, scent sprays, vacuum flasks and other vacuum vessels for personal or household use, wigs, false beards, eyebrows
– manufacture of miscellaneous articles: candles, tapers and the like; artificial flowers, fruit and foliage; jokes and novelties; hand sieves and hand riddles; tailors' dummies; burial coffins etc.
– manufacture of floral baskets, bouquets, wreaths and similar articles
– taxidermy activities

This class excludes:

– *manufacture of lighter wicks, see 13.96*
– *manufacture of workwear and service apparel (e.g. laboratory coats, work overalls, uniforms), see 14.12*
– *manufacture of paper novelties, see 17.29*

33   **Repair and installation of machinery and equipment**

This division includes the specialised repair of goods produced in the manufacturing sector with the aim of restoring machinery, equipment and other products to working order. The provision of general or routine maintenance (i.e. servicing) on such products to ensure they work efficiently and to prevent breakdown and unnecessary repairs is included.

This division does only include specialised repair and maintenance activities. A substantial amount of repair is also done by manufacturers of machinery, equipment and other goods, in which case the classification of units engaged in these repair and manufacturing activities is done according to the value added principle which would often assign these combined activities to the manufacture of the good. The same principle is applied for combined trade and repair.

The rebuilding or remanufacturing of machinery and equipment is considered a manufacturing activity and included in other divisions of this section.

Repair and maintenance of goods that are utilised as capital goods as well as consumer goods is typically classified as repair and maintenance of household goods (e.g. office and household furniture repair, see 95.24).

Also included in this division is the specialised installation of machinery. However, the installation of equipment that forms an integral part of buildings or similar structures, such as installation of electrical wiring, installation of escalators or installation of air-conditioning systems, is classified as construction.

This division excludes:

– *cleaning of industrial machinery, see 81.22*
– *repair and maintenance of computers and communications equipment, see 95.1*
– *repair and maintenance of household goods, see 95.2*

33.1   **Repair of fabricated metal products, machinery and equipment**

This group includes the specialised repair of goods produced in the manufacturing sector with the aim of restoring these metal products, machinery, equipment and other products to working order. The provision of general or routine maintenance (i.e. servicing) on such products to ensure they work efficiently and to prevent breakdown and unnecessary repairs is included.

This group excludes:

– *rebuilding or remanufacturing of machinery and equipment, see corresponding class in divisions 25–30*
– *cleaning of industrial machinery, see 81.22*
– *repair and maintenance of computers and communications equipment, see 95.1*
– *repair and maintenance of household goods, see 95.2*

33.11   **Repair of fabricated metal products**

This class includes the repair and maintenance of fabricated metal products of division 25.

This class includes:

– repair of metal tanks, reservoirs and containers
– repair and maintenance for pipes and pipelines

- mobile welding repair
- repair of steel shipping drums
- repair and maintenance of steam or other vapour generators
- repair and maintenance of auxiliary plant for use with steam generators:
  - condensers, economisers, superheaters, steam collectors and accumulators
- repair and maintenance of nuclear reactors, except isotope separators
- repair and maintenance of parts for marine or power boilers
- platework repair of central heating boilers and radiators
- repair and maintenance of firearms and ordnance (including repair of sporting and recreational guns)
- repair and maintenance of shopping carts

This class excludes:

- sharpening of blades and saws, see 33.12
- repair of central heating systems etc., see 43.22
- repair of mechanical locking devices, safes etc., see 80.20

33.12   Repair of machinery

This class includes the repair and maintenance of industrial machinery and equipment like sharpening or installing commercial and industrial machinery blades and saws; the provision of welding (e.g. automotive, general) repair services; the repair of agricultural and other heavy and industrial machinery and equipment (e.g. forklifts and other materials handling equipment, machine tools, commercial refrigeration equipment, construction equipment and mining machinery), including machinery and equipment of division 28.

This class includes:

- repair and maintenance of non-motor vehicle engines
- repair and maintenance of pumps, compressors and related equipment
- repair and maintenance of fluid power machinery
- repair of valves
- repair of gearing and driving elements
- repair and maintenance of industrial process furnaces
- repair and maintenance of lifting and handling equipment including lifts, elevators and moving walkways
- repair and maintenance of industrial refrigeration equipment and air purifying equipment
- repair and maintenance of commercial-type general purpose machinery
- repair of power-driven hand-tools
- repair and maintenance of metal cutting and metal forming machine tools and accessories
- repair and maintenance of other machine tools
- repair and maintenance of agricultural tractors
- repair and maintenance of agricultural machinery and forestry and logging machinery
- repair and maintenance of metallurgy machinery
- repair and maintenance of mining, construction, and oil and gas field machinery
- repair and maintenance of food, beverage, and tobacco processing machinery
- repair and maintenance of textile apparel, and leather production machinery
- repair and maintenance of papermaking machinery
- repair and maintenance of plastic and rubber processing machinery
- repair and maintenance of other special purpose machinery of division 28
- repair and maintenance of weighing equipment
- repair and maintenance of vending machines
- repair and maintenance of cash registers
- repair and maintenance of photocopy machines
- repair of calculators, electronic or not
- repair of typewriters

This class excludes:

- repair and maintenance of furnaces and other heating equipment, see 43.22
- installation, repair and maintenance of elevators and escalators in buildings and civil engineering structures, see 43.29
- repair of computers, see 95.11

33.13   Repair of electronic and optical equipment

This class includes the repair and maintenance of goods produced in groups 26.5, 26.6 and 26.7, except those that are considered household goods.

C

This class includes:

– repair and maintenance of the measuring, testing, navigating and control equipment of group 26.5, such as:
  ■ aircraft engine instruments
  ■ automotive emissions testing equipment
  ■ meteorological instruments
  ■ physical, electrical and chemical properties testing and inspection equipment
  ■ surveying instruments
  ■ radiation detection and monitoring instruments
– repair and maintenance of irradiation, electromedical and electrotherapeutic equipment of class 26.60, such as:
  ■ magnetic resonance imaging equipment
  ■ medical ultrasound equipment
  ■ pacemakers
  ■ hearing aids
  ■ electrocardiographs
  ■ electromedical endoscopic equipment
  ■ irradiation apparatus
– repair and maintenance of optical instruments and equipment of class 26.70, if the use is mainly commercial, such as:
  ■ binoculars
  ■ microscopes (except electron and proton microscopes)
  ■ telescopes
  ■ prisms and lenses (except ophthalmic)
  ■ photographic equipment

This class excludes:

– *repair and maintenance of photocopy machines, see 33.12*
– *repair and maintenance of computers and peripheral equipment, see 95.11*
– *repair and maintenance of computer projectors, see 95.11*
– *repair and maintenance of communication equipment, see 95.12*
– *repair and maintenance of commercial TV and video cameras, see 95.12*
– *repair of household-type video cameras, see 95.21*
– *repair of watches and clocks, see 95.25*

33.14   Repair of electrical equipment

This class includes the repair and maintenance of goods of division 27, except those in class group 27.5 (domestic appliances).

This class includes:

– repair and maintenance of power, distribution, and specialty transformers
– repair and maintenance of electric motors, generators, and motor generator sets
– repair and maintenance of switchgear and switchboard apparatus
– repair and maintenance of relays and industrial controls
– repair and maintenance of primary and storage batteries
– repair and maintenance of electric lighting equipment
– repair and maintenance of current-carrying wiring devices and non current-carrying wiring devices for wiring electrical circuits

This class excludes:

– *repair and maintenance of computers and peripheral computer equipment, see 95.11*
– *repair and maintenance of telecommunications equipment, see 95.12*
– *repair and maintenance of consumer electronics, see 95.21*
– *repair of watches and clocks, see 95.25*

33.15   Repair and maintenance of ships and boats

This class includes the repair and maintenance of ships and boats. However, the factory rebuilding or overhaul of ships is classified in division 30.

This class includes:

– repair and routine maintenance of ships
– repair and maintenance of pleasure boats

This class excludes:

- *factory conversion of ships, see 30.1*
- *repair of ship and boat engines, see 33.12*
- *ship scrapping, dismantling, see 38.31*

33.16   Repair and maintenance of aircraft and spacecraft

This class includes the repair and maintenance of aircraft and spacecraft.

This class includes:

- repair and maintenance of aircraft (except factory conversion, factory overhaul, factory rebuilding)
- repair and maintenance of aircraft engines

This class excludes:

- *factory overhaul and rebuilding of aircraft, see 30.30*

33.17   Repair and maintenance of other transport equipment

This class includes the repair and maintenance of other transport equipment of division 30, except motorcycles and bicycles.

This class includes:

- repair and maintenance of locomotives and railway cars (except factory rebuilding or factory conversion)
- repair of animal drawn buggies and wagons

This class excludes:

- *factory overhaul and rebuilding of locomotives and railway cars, see 30.20*
- *repair and maintenance of military fighting vehicles, see 30.40*
- *repair and maintenance of shopping carts, see 33.11*
- *repair and maintenance of railway engines, see 33.12*
- *repair and maintenance of motorcycles, see 45.40*
- *repair of bicycles, see 95.29*

33.19   Repair of other equipment

This class includes the repair and maintenance of equipment not covered in other groups of this division.

This class includes:

- repair of fishing nets, including mending
- repair or ropes, riggings, canvas and tarpaulins
- repair of fertiliser and chemical storage bags
- repair or reconditioning of wooden pallets, shipping drums or barrels, and similar items
- repair of pinball machines and other coin-operated games
- restoring of organs and other historical musical instruments

This class excludes:

- *repair of household and office type furniture, furniture restoration, see 95.24*
- *repair of bicycles, see 95.29*
- *repair and alteration of clothing, see 95.29*

33.2   Installation of industrial machinery and equipment

33.20   Installation of industrial machinery and equipment

This class includes the specialised installation of machinery. However, the installation of equipment that forms an integral part of buildings or similar structures, such as installation of escalators, electrical wiring, burglar alarm systems or air-conditioning systems, is classified as construction.

This class includes:

- installation of industrial machinery in industrial plant
- assembling of industrial process control equipment
- installation of other industrial equipment, e.g.:
  - communications equipment
  - mainframe and similar computers
  - irradiation and electromedical equipment etc.

- dismantling large-scale machinery and equipment
- activities of millwrights
- machine rigging
- installation of bowling alley equipment

This class excludes:

- *installation of elevators, escalators, automated doors, vacuum cleaning systems etc., see 43.29*
- *installation of doors, staircases, shop fittings, furniture etc., see 43.32*
- *installation (setting-up) of personal computers, see 62.09*

Section D　　**Electricity, Gas, Steam and Air Conditioning Supply**

This section includes the activity of providing electric power, natural gas, steam, hot water and the like through a permanent infrastructure (network) of lines, mains and pipes. The dimension of the network is not decisive; also included are the distribution of electricity, gas, steam, hot water and the like in industrial parks or residential buildings.

This section therefore includes the operation of electric and gas utilities, which generate, control and distribute electric power or gas. Also included is the provision of steam and air-conditioning supply.

This section excludes the operation of water and sewerage utilities, see 36, 37. This section also excludes the (typically long-distance) transport of gas through pipelines.

35　　**Electricity, gas, steam and air conditioning supply**

35.1　　**Electric power generation, transmission and distribution**

This group includes the generation of bulk electric power, transmission from generating facilities to distribution centres, and distribution to end users.

35.11　　Production of electricity

This class includes:

– operation of generation facilities that produce electric energy; including thermal, nuclear, hydroelectric, gas turbine, diesel and renewable

This class excludes:

– *production of electricity through incineration of waste, see 38.21*

35.12　　Transmission of electricity

This class includes:

– operation of transmission systems that convey the electricity from the generation facility to the distribution system

35.13　　Distribution of electricity

This class includes:

– operation of distribution systems (i.e., consisting of lines, poles, meters, and wiring) that convey electric power received from the generation facility or the transmission system to the final consumer

35.14　　Trade of electricity

This class includes:

– sale of electricity to the user
– activities of electric power brokers or agents that arrange the sale of electricity via power distribution systems operated by others
– operation of electricity and transmission capacity exchanges for electric power

35.2　　**Manufacture of gas; distribution of gaseous fuels through mains**

This group includes the manufacture of gas and the distribution of natural or synthetic gas to the consumer through a system of mains. Gas marketers or brokers, who arrange the sale of natural gas over distribution systems operated by others, are included.

The separate operation of gas pipelines, typically done over long distances, connecting producers with distributors of gas, or between urban centres, is excluded from this group and classified with other pipeline transport activities.

35.21　　Manufacture of gas

This class includes:

– production of gas for the purpose of gas supply by carbonation of coal, from by-products of agriculture or from waste
– manufacture of gaseous fuels with a specified calorific value, by purification, blending and other processes from gases of various types including natural gas

This class excludes:

– *production of crude natural gas, see 06.20*
– *operation of coke ovens, see 19.10*
– *manufacture of refined petroleum products, see 19.20*
– *manufacture of industrial gases, see 20.11*

35.22　　Distribution of gaseous fuels through mains

This class includes:

– distribution and supply of gaseous fuels of all kinds through a system of mains

This class excludes:

– *(long-distance) transportation of gases by pipelines, see 49.50*

D

35.23   **Trade of gas through mains**

This class includes:

– sale of gas to the user through mains
– activities of gas brokers or agents that arrange the sale of gas over gas distribution systems operated by others
– commodity and transport capacity exchanges for gaseous fuels

This class excludes:

– *wholesale of gaseous fuels, see 46.71*
– *retail sale of bottled gas, see 47.78*
– *direct selling of fuel, see 47.99*

35.3   **Steam and air conditioning supply**

35.30   **Steam and air conditioning supply**

This class includes:

– production, collection and distribution of steam and hot water for heating, power and other purposes
– production and distribution of cooled air
– production and distribution of chilled water for cooling purposes
– production of ice, for food and non-food (e.g. cooling) purposes

Section E   Water Supply; Sewerage, Waste Management and Remediation Activities

This section includes activities related to the management (including collection, treatment and disposal) of various forms of waste, such as solid or non-solid industrial or household waste, as well as contaminated sites. The output of the waste or sewage treatment process can either be disposed of or become an input into other production processes.

Activities of water supply are also grouped in this section, since they are often carried out in connection with, or by units also engaged in, the treatment of sewage.

36   Water collection, treatment and supply

This division includes the collection, treatment and distribution of water for domestic and industrial needs. Collection of water from various sources, as well as distribution by various means is included.

36.0   Water collection, treatment and supply

36.00   Water collection, treatment and supply

This class includes water collection, treatment and distribution activities for domestic and industrial needs. Collection of water from various sources, as well as distribution by various means is included.

The operation of irrigation canals is also included; however the provision of irrigation services through sprinklers, and similar agricultural support services, is not included.

This class includes:

– collection of water from rivers, lakes, wells etc.
– collection of rain water
– purification of water for water supply purposes
– treatment of water for industrial and other purposes
– desalting of sea or ground water to produce water as the principal product of interest
– distribution of water through mains, by trucks or other means
– operation of irrigation canals

This class excludes:

– *operation of irrigation equipment for agricultural purposes, see 01.61*
– *treatment of waste water in order to prevent pollution, see 37.00*
– *(long-distance) transport of water via pipelines, see 49.50*

37   Sewerage

This division includes the operation of sewer systems or sewage treatment facilities that collect, treat, and dispose of sewage.

37.0   Sewerage

37.00   Sewerage

This class includes:

– the operation of sewer systems or sewer treatment facilities
– collecting and transporting of human waste water from one or several users, as well as rain water by means of sewerage networks, collectors, tanks and other means of transport (sewage vehicles etc.)
– emptying and cleaning of cesspools and septic tanks, sinks and pits from sewage; servicing of chemical toilets
– treatment of waste water (including human and industrial waste water, water from swimming pools etc.) by means of physical, chemical and biological processes like dilution, screening, filtering, sedimentation etc.
– maintenance and cleaning of sewers and drains including sewer rodding

This class excludes:

– *decontamination of surface water and groundwater at the place of pollution, see 39.00*
– *cleaning and deblocking of drainpipes in buildings, see 43.22*

38   Waste collection, treatment and disposal activities; materials recovery

This division includes the collection, treatment, and disposal of waste materials. This also includes local hauling of waste materials and the operation of materials recovery facilities (i.e. those that sort recoverable materials from a waste stream).

38.1   Waste collection

This group includes the collection of waste from households and businesses by means of refuse bins, wheeled bins, containers, etc. It includes collection of non-hazardous and hazardous waste e.g. waste from households, used batteries, used cooking oils and fats, waste oil from ships and used oil from garages, as well as construction and demolition waste.

38.11   Collection of non-hazardous waste

This class includes:

- collection of non-hazardous solid waste (i.e. garbage) within a local area, such as collection of waste, which may include mixed recoverable materials, from households and businesses by means of refuse bins, wheeled bins, containers etc.
- collection of recyclable materials
- collection of refuse in litter-bins in public places

This class also includes:

- collection of construction and demolition waste
- collection and removal of debris such as brush and rubble
- collection of waste output of textile mills
- operation of waste transfer facilities for non-hazardous waste

This class excludes:

- *collection of hazardous waste, see 38.12*
- *operation of landfills for the disposal of non-hazardous waste, see 38.21*
- *operation of facilities where mixed recoverable materials such as paper, plastics, etc. are sorted into distinct categories, see 38.32*

38.12   Collection of hazardous waste

This class includes the collection of solid and non-solid hazardous waste, i.e. explosive, oxidising, flammable, toxic, irritant, carcinogenic, corrosive, infectious or other substances and preparations harmful to human health and the environment. It may also entail identification, treatment, packaging and labelling of waste for the purposes of transport.

This class includes:

- collection of hazardous waste, such as:
  - used oil from shipment or garages
  - bio-hazardous waste
  - nuclear waste
  - used batteries etc.
- operation of waste transfer stations for hazardous waste

This class excludes:

- *remediation and clean up of contaminated buildings, mine sites, soil, ground water, e.g. asbestos removal, see 39.00*

38.2   **Waste treatment and disposal**

This group includes the disposal and treatment prior to disposal of various forms of waste by different means, such as treatment of organic waste with the aim of disposal; treatment and disposal of toxic live or dead animals and other contaminated waste; treatment and disposal of transition radioactive waste from hospitals, etc.; dumping of refuse on land or in water; burial or ploughing-under of refuse; disposal of used goods such as refrigerators to eliminate harmful waste; disposal of waste by incineration or combustion. Energy recovery resulting from waste incineration processes is also included.

This group excludes:

- *treatment and disposal of waste water (see class 37.00).*
- *materials recovery, see 38.3*

38.21   Treatment and disposal of non-hazardous waste

This class includes the disposal and treatment prior to disposal of solid or non-solid non-hazardous waste:

- operation of landfills for the disposal of non-hazardous waste
- disposal of non-hazardous waste by combustion or incineration or other methods, with or without the resulting production of electricity or steam, compost, substitute fuels, biogas, ashes or other by-products for further use etc.
- treatment of organic waste for disposal

This class excludes:

- *incineration and combustion of hazardous waste, see 38.22*
- *operation of facilities where mixed recoverable materials such as paper, plastics, used beverage cans and metals are sorted into distinct categories, see 38.32*
- *decontamination, clean up of land, water; toxic material abatement, see 39.00*

38.22   Treatment and disposal of hazardous waste

This class includes the disposal and treatment prior to disposal of solid or non-solid hazardous waste, including waste that is explosive, oxidising, flammable, toxic, irritant, carcinogenic, corrosive, infectious and other substances and preparations harmful to human health and the environment.

This class includes:

- operation of facilities for treatment of hazardous waste
- treatment and disposal of toxic live or dead animals and other contaminated waste
- incineration of hazardous waste
- disposal of used goods such as refrigerators to eliminate harmful waste
- treatment, disposal and storage of radioactive nuclear waste including:
  - treatment and disposal of transition radioactive waste, i.e. decaying within the period of transport, from hospitals
  - encapsulation, preparation and other treatment of nuclear waste for storage

This class excludes:

- *reprocessing of nuclear fuels, see 20.13*
- *incineration of non-hazardous waste, see 38.21*
- *decontamination, clean up of land, water; toxic material abatement, see 39.00*

38.3   **Materials recovery**

38.31   **Dismantling of wrecks**

This class includes dismantling of wrecks of any type (automobiles, ships, computers, televisions and other equipment) for materials recovery.

This class excludes:

- *disposal of used goods such as refrigerators to eliminate harmful waste, see 38.22*
- *dismantling of automobiles, ships, computers, televisions and other equipment to obtain re-sell usable parts, see section G*

38.32   **Recovery of sorted materials**

This class includes the processing of metal and non-metal waste and scrap and other articles into secondary raw materials, usually involving a mechanical or chemical transformation process.

Also included is the recovery of materials from waste streams in the form of (1) separating and sorting recoverable materials from non-hazardous waste streams (i.e. garbage) or (2) the separating and sorting of mixed recoverable materials, such as paper, plastics, used beverage cans and metals, into distinct categories.

Examples of the mechanical or chemical transformation processes that are undertaken are:

- mechanical crushing of metal waste from used cars, washing machines, bikes etc.
- mechanical reduction of large iron pieces such as railway wagons
- shredding of metal waste, end-of-life vehicles etc.
- other methods of mechanical treatment such as cutting, pressing to reduce the volume
- reclaiming metals out of photographic waste, e.g. fixer solution or photographic films and paper
- reclaiming of rubber such as used tyres to produce secondary raw material
- sorting and pelleting of plastics to produce secondary raw material for tubes, flower pots, pallets and the like
- processing (cleaning, melting, grinding) of plastic or rubber waste to granulates
- crushing, cleaning and sorting of glass
- crushing, cleaning and sorting of other waste such as demolition waste to obtain secondary raw material
- processing of used cooking oils and fats into secondary raw materials
- processing of other food, beverage and tobacco waste and residual substances into secondary raw materials

This class excludes:

- *manufacture of new final products from (whether or not self-manufactured) secondary raw materials, such as spinning yarn from garnetted stock, making pulp from paper waste, retreading tyres or production of metal from metal scrap, see corresponding classes in section C (Manufacturing)*
- *reprocessing of nuclear fuels, see 20.13*
- *remelting ferrous waste and scrap, see 24.10*
- *materials recovery during waste combustion or incineration process, see 38.2*
- *treatment and disposal of non-hazardous waste, see 38.21*
- *treatment of organic waste for disposal, including production of compost, see 38.21*
- *energy recovery during non-hazardous waste incineration processes, see 38.21*
- *treatment and disposal of transition radioactive waste from hospitals etc., see 38.22*
- *treatment and disposal of toxic, contaminated waste, see 38.22*
- *wholesale of recoverable materials, see 46.77*

39   **Remediation activities and other waste management services**

This division includes the provision of remediation services, i.e. the cleanup of contaminated buildings and sites, soil, surface or ground water.

39.0   **Remediation activities and other waste management services**

39.00   Remediation activities and other waste management services

This class includes:

– decontamination of soils and groundwater at the place of pollution, either in situ or ex situ, using e.g. mechanical, chemical or biological methods
– decontamination of industrial plants or sites, including nuclear plants and sites
– decontamination and cleaning up of surface water following accidental pollution, e.g. through collection of pollutants or through application of chemicals
– cleaning up oil spills and other pollutions on land, in surface water, in ocean and seas, including coastal areas
– asbestos, lead paint, and other toxic material abatement
– other specialised pollution-control activities

This class excludes:

– *pest control in agriculture, see 01.61*
– *purification of water for water supply purposes, see 36.00*
– *treatment and disposal of non-hazardous waste, see 38.21*
– *treatment and disposal of hazardous waste, see 38.22*
– *outdoor sweeping and watering of streets etc., see 81.29*

Section F   Construction

This section includes general construction and specialised construction activities for buildings and civil engineering works. It includes new work, repair, additions and alterations, the erection of prefabricated buildings or structures on the site and also construction of a temporary nature.

General construction is the construction of entire dwellings, office buildings, stores and other public and utility buildings, farm buildings etc., or the construction of civil engineering works such as motorways, streets, bridges, tunnels, railways, airfields, harbours and other water projects, irrigation systems, sewerage systems, industrial facilities, pipelines and electric lines, sports facilities etc.

This work can be carried out on own account or on a fee or contract basis. Portions of the work and sometimes even the whole practical work can be subcontracted out. A unit that carries the overall responsibility for a construction project is classified here.

Also included is the repair of buildings and civil engineering works.

This section includes the complete construction of buildings (division 41), the complete construction of civil engineering works (division 42), as well as specialised construction activities, if carried out only as a part of the construction process (division 43).

The renting of construction equipment with operator is classified with the specific construction activity carried out with this equipment.

This section also includes the development of building projects for buildings or civil engineering works by bringing together financial, technical and physical means to realise the construction projects for later sale.

If these activities are carried out not for later sale of the construction projects, but for their operation (e.g. renting of space in these buildings, manufacturing activities in these plants), the unit would not be classified here, but according to its operational activity, i.e. real estate, manufacturing etc.

41   Construction of buildings

This division includes general construction of buildings of all kinds. It includes new work, repair, additions and alterations, the erection of pre-fabricated buildings or structures on the site and also construction of a temporary nature.

Included is the construction of entire dwellings, office buildings, stores and other public and utility buildings, farm buildings, etc.

41.1   Development of building projects

41.10   Development of building projects

This class includes:

– development of building projects for residential and non-residential buildings by bringing together financial, technical and physical means to realise the building projects for later sale

This class excludes:

– *construction of buildings, see 41.20*
– *architectural and engineering activities, see 71.1*
– *project management services related to building projects, see 71.1*

41.2   Construction of residential and non-residential buildings

This group includes the construction of complete residential or non-residential buildings, on own account for sale or on a fee or contract basis. Outsourcing parts or even the whole construction process is possible. If only specialised parts of the construction process are carried out, the activity is classified in division 43.

41.20   Construction of buildings

41.20/1   Construction of commercial buildings

This subclass includes:

– construction of all types of non-residential buildings:
  - buildings for industrial production, e.g. factories, workshops, assembly plants etc.
  - hospitals, primary, secondary and other schools, office buildings
  - hotels, stores, shopping malls, restaurants
  - airport buildings
  - indoor sports facilities
  - parking garages, including underground parking garages
  - warehouses
  - religious buildings
  - arts, cultural or leisure facilities buildings
– assembly and erection of prefabricated non-residential constructions on the site

This subclass excludes:

– *construction of industrial facilities, except buildings, see 42.99*
– *architectural and engineering activities, see 71.1*
– *project management for construction, see 71.1*

41.20/2 Construction of domestic buildings

This subclass includes:

– construction of all types of residential buildings:
  - single-family houses
  - multi-family buildings, including high-rise buildings
  - housing association and local authority housing

This subclass also includes:

– remodelling or renovating existing residential structures

This subclass excludes:

– *architectural and engineering activities, see 71.1*
– *project management for construction, see 71.1*

42 **Civil engineering**

This division includes general construction for civil engineering works. It includes new work, repair, additions and alterations, the erection of pre-fabricated structures on the site and also construction of temporary nature.

Included is the construction of heavy constructions such as motorways, streets, bridges, tunnels, railways, airfields, harbours and other water projects, irrigation systems, sewerage systems, industrial facilities, pipelines and electric lines, outdoor sports facilities, etc. This work can be carried out on own account or on a fee or contract basis. Portions of the work and sometimes even the whole practical work can be subcontracted out.

42.1 **Construction of roads and railways**

42.11 Construction of roads and motorways

This class includes:

– construction of motorways, streets, roads, other vehicular and pedestrian ways
– surface work on streets, roads, highways, bridges or tunnels:
  - asphalt paving of roads
  - road painting and other marking
  - installation of crash barriers, traffic signs and the like
– construction of airfield runways

This class excludes:

– *installation of street lighting and electrical signals, see 43.21*
– *architectural and engineering activities, see 71.1*
– *project management for construction, see 71.1*

42.12 Construction of railways and underground railways

This class includes:

– construction of railways and subways

This class excludes:

– *installation of lighting and electrical signals, see 43.21*
– *architectural and engineering activities, see 71.1*
– *project management for construction, see 71.1*

42.13 Construction of bridges and tunnels

This class includes:

– construction of bridges, including those for elevated highways
– construction of tunnels

This class excludes:

– *installation of lighting and electrical signals, see 43.21*
– *architectural and engineering activities, see 71.1*
– *project management for construction, see 71.1*

42.2　　　**Construction of utility projects**

42.21　　**Construction of utility projects for fluids**

This class includes the construction of distribution lines for transportation of fluids and related buildings and structures that are an integral part of these systems.

This class includes:

– construction of civil engineering constructions for:
  - long-distance and urban pipelines
  - water main and line construction
  - irrigation systems (canals)
  - reservoirs
– construction of:
  - sewer systems, including repair
  - sewage disposal plants
  - pumping stations

This class also includes:

– water well drilling

This class excludes:

– *project management activities related to civil engineering works, see 71.12*

42.22　　**Construction of utility projects for electricity and telecommunications**

This class includes the construction of distribution lines for electricity and telecommunications and related buildings and structures that are an integral part of these systems.

This class includes:

– construction of civil engineering constructions for:
  - long-distance and urban communication and power lines
  - power plants

This class excludes:

– *project management activities related to civil engineering works, see 71.12*

42.9　　　**Construction of other civil engineering projects**

42.91　　**Construction of water projects**

This class includes:

– construction of:
  - waterways, harbour and river works, pleasure ports (marinas), locks, etc.
  - dams and dykes
– dredging of waterways

This class excludes:

– *project management activities related to civil engineering works, see 71.12*

42.99　　**Construction of other civil engineering projects n.e.c.**

This class includes:

– construction of industrial facilities, except buildings, such as:
  - refineries
  - chemical plants
– construction work, other than buildings, such as:
  - outdoor sports facilities

This class also includes:

– land subdivision with land improvement (e.g. adding of roads, utility infrastructure etc.)

This class excludes:

– *installation of machinery and equipment, see 33.20*
– *land subdivision without land improvement, see 68.10*
– *project management activities related to civil engineering works, see 71.12*

43　　　　　　　**Specialised construction activities**

This division includes specialised construction activities (special trades), i.e. the construction, or preparation for construction, of parts of buildings and civil engineering works. These activities are usually specialised in one aspect common to different structures, requiring specialised skills or equipment, such as pile-driving, foundation work, carcass work, concrete work, brick laying, stone setting, scaffolding, roof covering, etc. The erection of steel structures is included provided that the parts are not produced by the same unit. Specialised construction activities are mostly carried out under subcontract, but especially in repair construction it is done directly for the owner of the property.

Also included are building finishing and building completion activities.

Included is the installation of all kind of utilities that make the construction function as such. These activities are usually performed at the site of the construction, although parts of the job may be carried out in a special shop. Included are activities such as plumbing, installation of heating and air-conditioning systems, antennas, alarm systems and other electrical work, sprinkler systems, elevators and escalators, etc. Also included are insulation work (water, heat, sound), sheet metal work, commercial refrigerating work, the installation of illumination and signalling systems for roads, railways, airports, harbours, etc. Repair of the above mentioned installations is also included.

Building completion activities encompass activities that contribute to the completion or finishing of a construction such as glazing, plastering, painting, floor and wall tiling or covering with other materials like parquet, carpets, wallpaper, etc., floor sanding, finish carpentry, acoustical work, cleaning of the exterior, etc. Repairs to the above mentioned completion or finishing work are also included.

The renting of equipment with operator is classified with the associated construction activity.

43.1　　　　　**Demolition and site preparation**

This group includes activities of preparing a site for subsequent construction activities, including the removal of previously existing structures.

43.11　　　　**Demolition**

This class includes:

– demolition or wrecking of buildings and other structures

43.12　　　　**Site preparation**

This class includes:

– clearing of building sites
– earth moving: excavation, landfill, levelling and grading of construction sites, trench digging, rock removal, blasting, etc.
– site preparation for mining:
　　■ overburden removal and other development and preparation of mineral properties and sites, except oil and gas sites

This class also includes:

– building site drainage
– drainage of agricultural or forestry land

This class excludes:

– *drilling of production oil or gas wells, see 06.10, 06.20*
– *decontamination of soil, see 39.00*
– *water well drilling, see 42.21*
– *shaft sinking, see 43.99*

43.13　　　　**Test drilling and boring**

This class includes:

– test drilling, test boring and core sampling for construction, geophysical, geological or similar purposes

This class excludes:

– *drilling of production oil or gas wells, see 06.10, 06.20*
– *test drilling and boring support services during mining activities, see 09.90*
– *water well drilling, see 42.21*
– *shaft sinking, see 43.99*
– *oil and gas field exploration, geophysical, geological and seismic surveying, see 71.12*

43.2　　　　**Electrical, plumbing and other construction installation activities**

This group includes installation activities that support the functioning of a building as such, including installation of electrical systems, plumbing (water, gas and sewage systems), heat and air-conditioning systems, elevators etc.

43.21　　　　Electrical installation

This class includes the installation of electrical systems in all kinds of buildings and civil engineering structures of electrical systems.

This class includes:

– installation of:
  - electrical wiring and fittings
  - telecommunications wiring
  - computer network and cable television wiring, including fibre optic
  - satellite dishes
  - lighting systems
  - fire alarms
  - burglar alarm systems
  - street lighting and electrical signals
  - airport runway lighting
  - electric solar energy collectors

This class also includes:

– connecting of electric appliances and household equipment, including baseboard heating

This class excludes:

– *construction of communications and power transmission lines, see 42.22*
– *monitoring and remote monitoring of electronic security systems, such as burglar alarms and fire alarms, including their installation and maintenance, see 80.20*

43.22　　　　Plumbing, heat and air-conditioning installation

This class includes the installation of plumbing, heating and air-conditioning systems, including additions, alterations, maintenance and repair.

This class includes:

– installation in buildings or other construction projects of:
  - heating systems (electric, gas and oil)
  - furnaces, cooling towers
  - non-electric solar energy collectors
  - plumbing and sanitary equipment
  - ventilation and air-conditioning equipment and ducts
  - gas fittings
  - steam piping
  - fire sprinkler systems
  - lawn sprinkler systems
– duct work installation

This class excludes:

– *installation of electric baseboard heating, see 43.21*

43.29　　　　Other construction installation

This class includes the installation of equipment other than electrical, plumbing, heating and air conditioning systems or industrial machinery in buildings and civil engineering structures.

This class includes:

– installation in buildings or other construction projects of:
  - elevators, escalators, including repair and maintenance
  - automated and revolving doors
  - lightning conductors
  - vacuum cleaning systems
  - thermal, sound or vibration insulation

This class excludes:

– *installation of industrial machinery, see 33.20*

43.3   **Building completion and finishing**

43.31   Plastering

This class includes:

– application in buildings or other construction projects of interior and exterior plaster or stucco, including related lathing materials

43.32   Joinery installation

This class includes:

– installation of doors (except automated and revolving), windows, door and window frames, of wood or other materials
– installation of fitted kitchens, built-in cupboards, staircases, shop fittings and the like
– interior completion such as ceilings, movable partitions, etc.

This class excludes:

– *installation of automated and revolving doors, see 43.29*

43.33   Floor and wall covering

This class includes:

– laying, tiling, hanging or fitting in buildings or other construction projects of:
  ■ ceramic, concrete or cut stone wall or floor tiles, ceramic stove fitting
  ■ parquet and other wooden floor coverings, wooden wall coverings
  ■ carpets and linoleum floor coverings, including of rubber or plastic
  ■ terrazzo, marble, granite or slate floor or wall coverings
  ■ wallpaper

43.34   Painting and glazing

43.34/1   Painting

This subclass includes:

– interior and exterior painting of buildings
– painting of civil engineering structures

43.34/2   Glazing

This subclass includes:

– installation of glass, mirrors, etc.

This subclass excludes:

– *installation of glazed window units and installation of unglazed window frames, see 43.32*

43.39   Other building completion and finishing

This class includes:

– cleaning of new buildings after construction
– other building completion and finishing work n.e.c.

This class excludes:

– *activities of interior decoration designers, see 74.10*
– *general interior cleaning of buildings and other structures, see 81.21*
– *specialised interior and exterior cleaning of buildings, see 81.22*

43.9   **Other specialised construction activities**

43.91   Roofing activities

This class includes:

– erection of roofs
– roof covering

This class excludes:

– *renting of construction machinery and equipment without operator, see 77.32*

43.99   Other specialised construction activities n.e.c.

43.99/1   Scaffold erection

This subclass includes:

– scaffolds and work platforms erected, dismantled and rented from the same unit

This subclass excludes:

– *renting of scaffolds without erection and dismantling, see 77.32*

43.99/9   Specialised construction activities (other than scaffold erection) n.e.c.

This subclass includes:

– construction activities specialising in one aspect common to different kind of structures, requiring specialised skill or equipment:
  - construction of foundations, including pile driving
  - damp proofing and water proofing works
  - de-humidification of buildings
  - shaft sinking
  - erection of steel elements
  - steel bending
  - bricklaying and stone setting
  - erection of chimneys and industrial ovens
  - work with specialist access requirements necessitating climbing skills and the use of related equipment, e.g. working at height on tall structures
– subsurface work
– construction of outdoor swimming pools
– steam cleaning, sand blasting and similar activities for building exteriors
– renting of cranes and other building equipment, which cannot be allocated to a specific construction type, with operator

This subclass excludes:

– *renting of construction machinery and equipment without operator, see 77.32*

F

Section G   **Wholesale and Retail Trade; Repair of Motor Vehicles and Motorcycles**

This section includes wholesale and retail sale (i.e. sale without transformation) of any type of goods, and the supply of services incidental to the sale of merchandise. Wholesaling and retailing are the final steps in the distribution of merchandise. Also included in this section are the repair of motor vehicles and motorcycles.

Sale without transformation is considered to include the usual operations (or manipulations) associated with trade, for example sorting, grading and assembling of goods, mixing (blending) of goods (for example sand), bottling (with or without preceding bottle cleaning), packing, breaking bulk and repacking for distribution in smaller lots, storage (whether or not frozen or chilled).

Division 45 includes all activities related to the sale and repair of motor vehicles and motorcycles, while divisions 46 and 47 include all other sale activities. The distinction between division 46 (wholesale) and division 47 (retail sale) is based on the predominant type of customer.

Wholesale is the resale (sale without transformation) of new and used goods to retailers, business to business trade such as to industrial, commercial, institutional or professional users, or resale to other wholesalers, or involves acting as an agent or broker in buying merchandise for, or selling merchandise to, such persons or companies. The principal types of businesses included are merchant wholesalers, i.e. wholesalers who take title to the goods they sell, such as wholesale merchants or jobbers, industrial distributors, exporters, importers, and cooperative buying associations, sales branches and sales offices (but not retail stores) that are maintained by manufacturing or mining units apart from their plants or mines for the purpose of marketing their products and that do not merely take orders to be filled by direct shipments from the plants or mines. Also included are merchandise and commodity brokers, commission merchants and agents and assemblers, buyers and cooperative associations engaged in the marketing of farm products.

Wholesalers frequently physically assemble, sort and grade goods in large lots, break bulk, repack and redistribute in smaller lots, for example pharmaceuticals; store, refrigerate, deliver and install goods, engage in sales promotion for their customers and label design.

Retailing is the resale (sale without transformation) of new and used goods mainly to the general public for personal or household consumption or utilisation, by shops, department stores, stalls, mail-order houses, door-to-door sales persons, hawkers, consumer cooperatives, auction houses etc. Most retailers take title to the goods they sell, but some act as agents for a principal and sell either on consignment or on a commission basis.

45   **Wholesale and retail trade and repair of motor vehicles and motorcycles**

This division includes all activities (except manufacture and renting) related to motor vehicles and motorcycles, including lorries and trucks, such as the wholesale and retail sale of new and second-hand vehicles, the repair and maintenance of vehicles and the wholesale and retail sale of parts and accessories for motor vehicles and motorcycles. Also included are activities of commission agents involved in wholesale or retail sale of vehicles, wholesale car auctions and wholesale over the Internet.

This division also includes activities such as washing, polishing of vehicles etc.

This division does not include the retail sale of automotive fuel and lubricating or cooling products or the renting of motor vehicles or motorcycles.

45.1   **Sale of motor vehicles**

45.11   **Sale of cars and light motor vehicles**

45.11/1   **Sale of new cars and light motor vehicles**

This subclass includes:

– wholesale and retail sale of new vehicles:
- passenger motor vehicles, including specialised passenger motor vehicles such as ambulances and minibuses, etc. (of a weight not exceeding 3.5 tonnes)

This subclass also includes:

– wholesale and retail sale of off-road motor vehicles (of a weight not exceeding 3.5 tonnes)

This subclass excludes:

– *wholesale and retail sale of parts and accessories for motor vehicles, see 45.3*
– *renting of motor vehicles with driver, see 49.3*
– *renting of motor vehicles without driver, see 77.1*

45.11/2   **Sale of used cars and light motor vehicles**

This subclass includes:

– wholesale and retail sale of used vehicles:
- passenger motor vehicles, including specialised passenger motor vehicles such as ambulances and minibuses, etc. (of a weight not exceeding 3.5 tonnes)

This subclass also includes:

– wholesale and retail sale of off-road motor vehicles (of a weight not exceeding 3.5 tonnes)

This subclass excludes:

– *wholesale and retail sale of parts and accessories for motor vehicles, see 45.3*
– *renting of motor vehicles with driver, see 49.3*
– *renting of motor vehicles without driver, see 77.1*

45.19   Sale of other motor vehicles

This class includes:

– wholesale and retail sale of new and used vehicles:
  - lorries, trailers and semi-trailers
  - camping vehicles such as caravans and motor homes

This class also includes:

– wholesale and retail sale of off-road motor vehicles (of a weight exceeding 3.5 tonnes)

This class excludes:

– *wholesale and retail sale of parts and accessories for motor vehicles, see 45.3*
– *renting of trucks with driver, see 49.41*
– *renting of trucks without driver, see 77.12*

45.2   Maintenance and repair of motor vehicles

45.20   Maintenance and repair of motor vehicles

This class includes:

– maintenance and repair of motor vehicles:
  - mechanical repairs
  - electrical repairs
  - electronic injection systems repair
  - ordinary servicing
  - bodywork repair
  - repair of motor vehicle parts
  - washing, polishing, etc.
  - spraying and painting
  - repair of screens and windows
  - repair of motor vehicle seats
– tyre and tube repair, fitting or replacement
– anti-rust treatment
– installation of parts and accessories not as part of the manufacturing process

This class excludes:

– *retreading and rebuilding of tyres, see 22.11*

45.3   Sale of motor vehicle parts and accessories

This group includes wholesale and retail trade of all kinds of parts, components, supplies, tools and accessories for motor vehicles, such as:

  - rubber tyres and inner tubes for tyres
  - spark plugs, batteries, lighting equipment and electrical parts

45.31   Wholesale trade of motor vehicle parts and accessories

45.32   Retail trade of motor vehicle parts and accessories

This class excludes:

– *retail sale of automotive fuel, see 47.30*

45.4   Sale, maintenance and repair of motorcycles and related parts and accessories

45.40   Sale, maintenance and repair of motorcycles and related parts and accessories

This class includes:

– wholesale and retail sale of motorcycles, including mopeds

- wholesale and retail sale of parts and accessories for motorcycles (including by commission agents and mail order houses)
- maintenance and repair of motorcycles

This class excludes:

- *wholesale of bicycles and related parts and accessories, see 46.49*
- *retail sale of bicycles and related parts and accessories, see 47.64*
- *renting of motorcycles, see 77.39*
- *repair and maintenance of bicycles, see 95.29*

46   **Wholesale trade, except of motor vehicles and motorcycles**

This division includes wholesale trade on own account or on a fee or contract basis (commission trade) related to domestic wholesale trade as well as international wholesale trade (import/export).

This division excludes:

- *wholesale of motor vehicles, caravans and motorcycles, see 45.1, 45.4*
- *wholesale of motor vehicle accessories, see 45.31, 45.40*
- *renting and leasing of goods, see division 77*
- *packing of solid goods and bottling of liquid or gaseous goods, including blending and filtering for third parties, see 82.92*

46.1   **Wholesale on a fee or contract basis**

This group includes:

- activities of commission agents, commodity brokers and all other wholesalers who trade on behalf and on the account of others
- activities of those involved in bringing sellers and buyers together or undertaking commercial transactions on behalf of a principal, including on the Internet.

This group also includes:

- activities of wholesale auctioneering houses, including Internet wholesale auctions

46.11   **Agents involved in the sale of agricultural raw materials, live animals, textile raw materials and semi-finished goods**

This class excludes:

- *wholesale trade in own name, see 46.2 to 46.9*
- *retail sale by non-store commission agents, see 47.99*

46.12   **Agents involved in the sale of fuels, ores, metals and industrial chemicals**

This class includes agents involved in the sale of:

- fuels, ores, metals and industrial chemicals, including fertilisers

This class excludes:

- *wholesale trade in own name, see 46.2 to 46.9*
- *retail sale by non-store commission agents, see 47.99*

46.13   **Agents involved in the sale of timber and building materials**

This class excludes:

- *wholesale trade in own name, see 46.2 to 46.9*
- *retail sale by non-store commission agents, see 47.99*

46.14   **Agents involved in the sale of machinery, industrial equipment, ships and aircraft**

This class includes agents involved in the sale of:

- machinery, including office machinery and computers, industrial equipment, ships and aircraft

This class excludes:

- *activities of commission agents for motor vehicles, see 45.1*
- *auctions of motor vehicles, see 45.1*
- *wholesale trade in own name, see 46.2 to 46.9*
- *retail sale by non-store commission agents, see 47.99*

46.15   **Agents involved in the sale of furniture, household goods, hardware and ironmongery**

This class excludes:

- *wholesale trade in own name, see 46.2 to 46.9*
- *retail sale by non-store commission agents, see 47.99*

46.16   Agents involved in the sale of textiles, clothing, fur, footwear and leather goods

This class excludes:

– *wholesale trade in own name, see 46.2 to 46.9*
– *retail sale by non-store commission agents, see 47.99*

46.17   Agents involved in the sale of food, beverages and tobacco

This class excludes:

– *wholesale trade in own name, see 46.2 to 46.9*
– *retail sale by non-store commission agents, see 47.99*

46.18   Agents specialised in the sale of other particular products

This class excludes:

– *wholesale trade in own name, see 46.2 to 46.9*
– *retail sale by non-store commission agents, see 47.99*
– *activities of insurance agents, see 66.22*
– *activities of real estate agents, see 68.31*

46.19   Agents involved in the sale of a variety of goods

This class excludes:

– *wholesale trade in own name, see 46.2 to 46.9*
– *retail sale by non-store commission agents, see 47.99*

46.2   Wholesale of agricultural raw materials and live animals

46.21   Wholesale of grain, unmanufactured tobacco, seeds and animal feeds

This class includes:

– wholesale of grains and seeds
– wholesale of oleaginous fruits
– wholesale of unmanufactured tobacco
– wholesale of animal feeds and agricultural raw material n.e.c.

This class excludes:

– *wholesale of textile fibres, see 46.76*

46.22   Wholesale of flowers and plants

This class includes:

– wholesale of flowers, plants and bulbs

46.23   Wholesale of live animals

46.24   Wholesale of hides, skins and leather

46.3   Wholesale of food, beverages and tobacco

46.31   Wholesale of fruit and vegetables

This class includes:

– wholesale of fresh fruit and vegetables
– wholesale of preserved fruit and vegetables

46.32   Wholesale of meat and meat products

46.33   Wholesale of dairy products, eggs and edible oils and fats

This class includes:

– wholesale of dairy products
– wholesale of eggs and egg products
– wholesale of edible oils and fats of animal or vegetable origin

46.34   Wholesale of alcoholic and other beverages

46.34/1   Wholesale of fruit and vegetable juices, mineral waters and soft drinks

This subclass includes:

– wholesale of non-alcoholic beverages

159

G

46.34/2   Wholesale of wine, beer, spirits and other alcoholic beverages

This subclass includes:

– wholesale of alcoholic beverages

This subclass also includes:

– buying of wine in bulk and bottling without transformation

This subclass excludes:

– *blending of wine or distilled spirits, see 11.01, 11.02*

46.35   Wholesale of tobacco products

46.36   Wholesale of sugar and chocolate and sugar confectionery

This class includes:

– wholesale of sugar, chocolate and sugar confectionery
– wholesale of bakery products

46.37   Wholesale of coffee, tea, cocoa and spices

46.38   Wholesale of other food, including fish, crustaceans and molluscs

This class also includes:

– wholesale of feed for pet animals

46.39   Non-specialised wholesale of food, beverages and tobacco

46.4   **Wholesale of household goods**

This group includes the wholesale of household goods, including textiles.

46.41   Wholesale of textiles

This class includes:

– wholesale of yarn
– wholesale of fabrics
– wholesale of household linen etc.
– wholesale of haberdashery: needles, sewing thread etc.

This class excludes:

– *wholesale of textile fibres, see 46.76*

46.42   Wholesale of clothing and footwear

This class includes:

– wholesale of clothing, including sports clothes
– wholesale of clothing accessories such as gloves, ties and braces
– wholesale of footwear
– wholesale of fur articles
– wholesale of umbrellas

This class excludes:

– *wholesale of jewellery, see 46.48*
– *wholesale of leather goods, see 46.49*
– *wholesale of special sports equipment footwear such as ski boots, see 46.49*

46.43   Wholesale of electrical household appliances

46.43/1   Wholesale of gramophone records, audio tapes, compact discs and video tapes and of the equipment on which these are played

This subclass includes:

– wholesale of recorded audio and video tapes, CDs, DVDs and the equipment on which these are played

This subclass excludes:

– *wholesale of blank audio and video tapes, CDs, DVDs, see 46.52*

46.43/9   Wholesale of radio and television goods and of electrical household appliances (other than of gramophone records, audio tapes, compact discs and video tapes and the equipment on which these are played) n.e.c.

This subclass includes:

- wholesale of electrical household appliances
- wholesale of radio and television equipment
- wholesale of photographic and optical goods
- wholesale of electrical heating appliances

This subclass excludes:

- *wholesale of sewing machines, see 46.64*

46.44   **Wholesale of china and glassware and cleaning materials**

This class includes:

- wholesale of china and glassware
- wholesale of cleaning materials

46.45   **Wholesale of perfume and cosmetics**

This class includes:

- wholesale of perfumeries, cosmetics and soaps

46.46   **Wholesale of pharmaceutical goods**

This class includes:

- wholesale of pharmaceutical and medical goods

46.47   **Wholesale of furniture, carpets and lighting equipment**

This class includes:

- wholesale of household furniture
- wholesale of carpets
- wholesale of lighting equipment

This class excludes:

- *wholesale oåf office furniture, see 46.65*

46.48   **Wholesale of watches and jewellery**

46.49   **Wholesale of other household goods**

46.49/1   **Wholesale of musical instruments**

46.49/9   **Wholesale of household goods (other than musical instruments) n.e.c.**

This subclass includes:

- wholesale of woodenware, wickerwork and corkware etc.
- wholesale of bicycles and their parts and accessories
- wholesale of stationery, books, magazines and newspapers
- wholesale of leather goods and travel accessories
- wholesale of games and toys
- wholesale of sports goods, including special sports footwear such as ski boots

46.5   **Wholesale of information and communication equipment**

This group includes the wholesale of information and communications technology (ICT) equipment, i.e. computers, telecommunications equipment and parts.

46.51   **Wholesale of computers, computer peripheral equipment and software**

This class includes:

- wholesale of computers and computer peripheral equipment
- wholesale of software

This class excludes:

- *wholesale of electronic parts, see 46.52*
- *wholesale of office machinery and equipment, (except computers and peripheral equipment), see 46.66*

46.52   **Wholesale of electronic and telecommunications equipment and parts**

This class includes:

- wholesale of electronic valves and tubes

– wholesale of semi-conductor devices
– wholesale of microchips and integrated circuits
– wholesale of printed circuits
– wholesale of blank audio and video tapes and diskettes, magnetic and optical disks (CDs, DVDs)
– wholesale of telephone and communications equipment

This class excludes:

– *wholesale of recorded audio and video tapes, CDs, DVDs, see 46.43*
– *wholesale of computers and computer peripheral equipment, see 46.51*

46.6   **Wholesale of other machinery, equipment and supplies**

This group includes the wholesale of specialised machinery, equipment and supplies for all kinds of industries and general purpose machinery.

46.61   **Wholesale of agricultural machinery, equipment and supplies**

This class includes:

– wholesale of agricultural machinery and equipment:
   - ploughs, manure spreaders, seeders
   - harvesters
   - threshers
   - milking machines
   - poultry-keeping machines, bee-keeping machines
   - tractors used in agriculture and forestry

This class also includes:

– lawn mowers however operated

46.62   **Wholesale of machine tools**

This class includes:

– wholesale of machine tools of any type and for any material

This class also includes:

– wholesale of computer-controlled machine tools

46.63   **Wholesale of mining, construction and civil engineering machinery**

46.64   **Wholesale of machinery for the textile industry and of sewing and knitting machines**

This class also includes:

– wholesale of computer-controlled machinery for the textile industry and of computer-controlled sewing and knitting machines

46.65   **Wholesale of office furniture**

This class includes:

– wholesale trade services related to:
   - goods classified in 31.01 (Manufacture of office and shop furniture)

46.66   **Wholesale of other office machinery and equipment**

This class includes:

– wholesale of office machinery and equipment, except computers and computer peripheral equipment

This class excludes:

– *wholesale of computers and peripheral equipment, see 46.51*
– *wholesale of electronic parts and telephone and communications equipment, see 46.52*

46.69   **Wholesale of other machinery and equipment**

This class includes:

– wholesale of transport equipment except motor vehicles, motorcycles and bicycles
– wholesale of production-line robots
– wholesale of wires and switches and other installation equipment for industrial use
– wholesale of other electrical material such as electrical motors, transformers

- wholesale of other machinery n.e.c. for use in industry (except mining, construction, civil engineering and textile industry), trade and navigation and other services

This class also includes:

- wholesale of measuring instruments and equipment

This class excludes:

- *wholesale of motor vehicles, trailers and caravans, see 45.1*
- *wholesale of motor vehicle parts, see 45.31*
- *wholesale of motorcycles, see 45.40*
- *wholesale of bicycles, see 46.49*

46.7　Other specialised wholesale

This group includes other specialised wholesale activities not classified in other groups of this division. This includes the wholesale of intermediate products, except agricultural, typically not for household use.

46.71　Wholesale of solid, liquid and gaseous fuels and related products

This class includes:

- wholesale of fuels, greases, lubricants, oils such as:
    - charcoal, coal, coke, fuel wood, naphtha
    - crude petroleum, crude oil, diesel fuel, gasoline, fuel oil, heating oil, kerosene
    - liquefied petroleum gases, butane and propane gas
    - lubricating oils and greases, refined petroleum products

46.71/1　Wholesale of petroleum and petroleum products

This subclass includes:

- wholesale of automotive fuels

46.71/9　Wholesale of fuels and related products (other than petroleum and petroleum products)

This subclass includes:

- wholesale of greases, lubricants, oils, etc.

46.72　Wholesale of metals and metal ores

This class includes:

- wholesale of ferrous and non-ferrous metal ores
- wholesale of ferrous and non-ferrous metals in primary forms
- wholesale of ferrous and non-ferrous semi-finished metal products n.e.c.
- wholesale of gold and other precious metals

This class excludes:

- *wholesale of metal scrap, see 46.77*

46.73　Wholesale of wood, construction materials and sanitary equipment

This class includes:

- wholesale of wood in the rough
- wholesale of products of primary processing of wood
- wholesale of paint and varnish
- wholesale of construction materials:
    - sand, gravel
- wholesale of wallpaper and floor coverings
- wholesale of flat glass
- wholesale of sanitary equipment:
    - baths, washbasins, toilets and other sanitary porcelain
- wholesale of prefabricated buildings

46.74　Wholesale of hardware, plumbing and heating equipment and supplies

This class includes:

- wholesale of hardware and locks
- wholesale of fittings and fixtures

  – wholesale of hot water heaters
  – wholesale of sanitary installation equipment:
    ■ tubes, pipes, fittings, taps, T-pieces, connections, rubber pipes etc.
  – wholesale of tools such as hammers, saws, screwdrivers and other hand tools

46.75   **Wholesale of chemical products**

This class includes:

  – wholesale of industrial chemicals:
    ■ aniline, printing ink, essential oils, industrial gases, chemical glues, colouring matter, synthetic resin, methanol, paraffin, scents and flavourings, soda, industrial salt, acids and sulphurs, starch derivates etc.
  – wholesale of fertilisers and agrochemical products

46.76   **Wholesale of other intermediate products**

This class includes:

  – wholesale of plastic materials in primary forms
  – wholesale of rubber
  – wholesale of textile fibres etc.
  – wholesale of paper in bulk
  – wholesale of precious stones

46.77   **Wholesale of waste and scrap**

This class includes:

  – wholesale of metal and non-metal waste and scrap and materials for recycling, including collecting, sorting, separating, stripping of used goods such as cars in order to obtain reusable parts, packing and repacking, storage and delivery, but without a real transformation process. Additionally, the purchased and sold waste has a remaining value.

This class also includes:

  – dismantling of automobiles, computers, televisions and other equipment to obtain and re-sell usable parts

This class excludes:

  – *collection of household and industrial waste, see 38.1*
  – *treatment of waste, not for a further use in an industrial manufacturing process, but with the aim of disposal, see 38.2*
  – *processing of waste and scrap and other articles into secondary raw material when a real transformation process is required (the resulting secondary raw material is fit for direct use in an industrial manufacturing process, but is not a final product), see 38.3*
  – *dismantling of automobiles, computers, televisions and other equipment for materials recovery, see 38.31*
  – *ship-breaking, see 38.31*
  – *shredding of cars by means of a mechanical process, see 38.32*
  – *retail sale of second-hand goods, see 47.79*

46.9   **Non-specialised wholesale trade**

46.90   Non-specialised wholesale trade

This class includes:

  – wholesale of a variety of goods without any particular specialisation

47   **Retail trade, except of motor vehicles and motorcycles**

This division includes the resale (sale without transformation) of new and used goods mainly to the general public for personal or household consumption or utilisation, by shops, department stores, stalls, mail-order houses, door-to-door sales persons, hawkers, consumer cooperatives etc.

Retail trade is classified first by type of sale outlet (retail trade in stores: groups 47.1 to 47.7; retail trade not in stores: groups 47.8 and 47.9). Retail trade in stores includes the retail sale of used goods (class 47.79). For retail sale in stores, there exists a further distinction between specialised retail sale (groups 47.2 to 47.7) and non-specialised retail sale (group 47.1). The above groups are further subdivided by the range of products sold. Sale not via stores is subdivided according to the forms of trade, such as retail sale via stalls and markets (group 47.8) and other non-store retail sale, e.g. mail order, door-to-door, by vending machines etc. (group 47.9).

The goods sold in this division are limited to goods usually referred to as consumer goods or retail goods. Therefore goods not normally entering the retail trade, such as cereal grains, ores, industrial machinery etc. are excluded.

This division also includes units engaged primarily in selling to the general public, from displayed merchandise, products such as personal computers, stationery, paint or timber, although these products may not be for personal or household use.

Handling that is customary in trade does not affect the basic character of the merchandise and may include, for example, sorting, separating, mixing and packaging.

This division also includes retail sale by commission agents and activities of retail auctioning houses.

This division excludes:

- *sale of farmers' products by farmers, see division 01*
- *manufacture and sale of goods, which is generally classified as manufacturing in divisions 10-32*
- *sale of motor vehicles, motorcycles and their parts, see division 45*
- *trade in cereal grains, ores, crude petroleum, industrial chemicals, iron and steel and industrial machinery and equipment, see division 46*
- *sale of food and drinks for consumption on the premises and sale of takeaway food, see division 56*
- *renting of personal and household goods to the general public, see group 77.2*

47.1   Retail sale in non-specialised stores

This group includes the retail sale of a variety of product lines in the same unit (non-specialised stores), such as supermarkets or department stores.

47.11   Retail sale in non-specialised stores with food, beverages or tobacco predominating

This class includes:

- retail sale of a large variety of goods of which, however, food products, beverages or tobacco should be predominant:
  - activities of general stores that have, apart from their main sales of food products, beverages or tobacco, several other lines of merchandise such as wearing apparel, furniture, appliances, hardware, cosmetics etc.

47.19   Other retail sale in non-specialised stores

This class includes:

- retail sale of a large variety of goods of which food products, beverages or tobacco are not predominant
- activities of department stores carrying a general line of merchandise, including wearing apparel, furniture, appliances, hardware, cosmetics, jewellery, toys, sports goods etc.

47.2   Retail sale of food, beverages and tobacco in specialised stores

47.21   Retail sale of fruit and vegetables in specialised stores

This class includes:

- retail sale of fresh fruit and vegetables
- retail sale of prepared and preserved fruits and vegetables

47.22   Retail sale of meat and meat products in specialised stores

This class includes:

- retail sale of meat and meat products (including poultry)

47.23   Retail sale of fish, crustaceans and molluscs in specialised stores

This class includes:

- retail sale of fish, other seafood and products thereof

47.24   Retail sale of bread, cakes, flour confectionery and sugar confectionery in specialised stores

47.25   Retail sale of beverages in specialised stores

This class includes:

- retail sale of beverages (not for consumption on the premises):
  - alcoholic beverages
  - non-alcoholic beverages

47.26   Retail sale of tobacco products in specialised stores

This class includes:

- retail sale of tobacco
- retail sale of tobacco products

47.29   Other retail sale of food in specialised stores

This class includes:

- retail sale of dairy products and eggs
- retail sale of other food products n.e.c.

47.3   **Retail sale of automotive fuel in specialised stores**

47.30   **Retail sale of automotive fuel in specialised stores**

This class includes:

– retail sale of fuel for motor vehicles and motorcycles

This class also includes:

– retail sale of lubricating products and cooling products for motor vehicles

This class excludes:

– *wholesale of fuels, see 46.71*
– *retail sale of liquefied petroleum gas for cooking or heating, see 47.78*

G

47.4   **Retail sale of information and communication equipment in specialised stores**

This group includes the retail sale of information and communications technology (ICT) equipment, such as computers and peripheral equipment, telecommunications equipment and consumer electronics, by specialised stores.

47.41   **Retail sale of computers, peripheral units and software in specialised stores**

This class includes:

– retail sale of computers
– retail sale of computer peripheral equipment
– retail sale of video game consoles
– retail sale of non-customised software, including video games

This class excludes:

– *retail sale of blank tapes and disks, see 47.63*

47.42   **Retail sale of telecommunications equipment in specialised stores**

47.42/1   **Retail sale of mobile telephones in specialised stores**

This subclass includes:

– specialised retail sale of mobile telephones
– specialised retail sale of mobile phones for motor vehicles
– installation of mobile phones in motor vehicles

This subclass excludes:

– *mobile telephone services, see 61*
– *paging services, see 61*

47.42/9   **Retail sale of telecommunications equipment (other than mobile telephones) n.e.c., in specialised stores**

47.43   **Retail sale of audio and video equipment in specialised stores**

This class includes:

– retail sale of radio and television equipment
– retail sale of audio and video equipment
– retail sale of CD, DVD etc. players and recorders

47.5   **Retail sale of other household equipment in specialised stores**

This group includes the retail sale of household equipment, such as textiles, hardware, carpets, electrical appliances or furniture, in specialised stores.

47.51   **Retail sale of textiles in specialised stores**

This class includes:

– retail sale of fabrics
– retail sale of knitting yarn
– retail sale of basic materials for rug, tapestry or embroidery making
– retail sale of textiles
– retail sale of haberdashery: needles, sewing thread etc.

This class excludes:

– *retail sale of clothing, see 47.71*

47.52　　Retail sale of hardware, paints and glass in specialised stores

This class includes:

- retail sale of hardware
- retail sale of paints, varnishes and lacquers
- retail sale of flat glass
- retail sale of other building material such as bricks, wood, sanitary equipment
- retail sale of do-it-yourself material and equipment

This class also includes:

- retail sale of lawnmowers, however operated
- retail sale of saunas

47.53　　Retail sale of carpets, rugs, wall and floor coverings in specialised stores

This class includes:

- retail sale of carpets and rugs
- retail sale of curtains and net curtains
- retail sale of wallpaper and floor coverings

This class excludes:

- *retail sale of cork floor tiles, see 47.52*

47.54　　Retail sale of electrical household appliances in specialised stores

This class excludes:

- *retail sale of audio and video equipment, see 47.43*

47.59　　Retail sale of furniture, lighting equipment and other household articles in specialised stores

47.59/1　　Retail sale of musical instruments and scores in specialised stores

This subclass includes:

- retail sale of musical instruments and scores

47.59/9　　Retail sale of furniture, lighting equipment and other household articles (other than musical instruments) n.e.c., in specialised stores

This subclass includes:

- retail sale of household furniture
- retail sale of articles for lighting
- retail sale of household utensils and cutlery, crockery, glassware, china and pottery
- retail sale of wooden, cork and wickerwork goods
- retail sale of non-electrical household appliances
- retail sale of electrical security alarm systems, such as locking devices, safes, and vaults, without installation or maintenance services
- retail sale of household articles and equipment n.e.c.

This subclass excludes:

- *retail sale of antiques, see 47.79*

47.6　　**Retail sale of cultural and recreation goods in specialised stores**

This group includes the retail sale in specialised stores of cultural and recreation goods, such as books, newspapers, music and video recordings, sporting equipment, games and toys.

47.61　　Retail sale of books in specialised stores

This class includes:

- retail sale of books of all kinds

This class excludes:

- *retail sale of second-hand or antique books, see 47.79*

47.62　　Retail sale of newspapers and stationery in specialised stores

This class also includes:

- retail sale of office supplies such as pens, pencils, paper etc.

47.63   **Retail sale of music and video recordings in specialised stores**

This class includes:

- retail sale of musical records, audio tapes, compact discs and cassettes
- retail sale of video tapes and DVDs

This class also includes:

- retail sale of blank tapes and discs

47.64   **Retail sale of sporting equipment in specialised stores**

This class includes:

- retail sale of sports goods, fishing gear, camping goods, boats and bicycles

47.65   **Retail sale of games and toys in specialised stores**

This class includes:

- retail sale of games and toys, made of all materials

This class excludes:

- *retail sale of video game consoles, see 47.41*
- *retail sale of non-customised software, including video games, see 47.41*

47.7   **Retail sale of other goods in specialised stores**

This group includes sale in specialised stores carrying a particular line of products not included in other parts of the classification, such as clothing, footwear and leather articles, pharmaceutical and medical goods, watches, souvenirs, cleaning materials, weapons, flowers and pets and others. Also included is the retail sale of used goods in specialised stores.

47.71   **Retail sale of clothing in specialised stores**

This class includes:

- retail sale of articles of clothing
- retail sale of articles of fur
- retail sale of clothing accessories such as gloves, ties, braces etc.

This class excludes:

- *retail sale of textiles, see 47.51*

47.72   **Retail sale of footwear and leather goods in specialised stores**

47.72/1   **Retail sale of footwear in specialised stores**

This subclass includes:

- retail sale of footwear

This subclass excludes:

- *retail sale of special sports equipment footwear such as ski boots, see 47.64*

47.72/2   **Retail sale of leather goods in specialised stores**

This subclass includes:

- retail sale of leather goods
- retail sale of travel accessories of leather and leather substitutes

47.73   **Dispensing chemist in specialised stores**

This class includes:

- retail sale of pharmaceuticals

47.74   **Retail sale of medical and orthopaedic goods in specialised stores**

47.74/1   **Retail sale of hearing aids in specialised stores**

47.74/9   **Retail sale of medical and orthopaedic goods (other than hearing aids) n.e.c., in specialised stores**

47.75   **Retail sale of cosmetic and toilet articles in specialised stores**

This class includes:

- retail sale of perfumery, cosmetic and toilet articles

47.76   **Retail sale of flowers, plants, seeds, fertilisers, pet animals and pet food in specialised stores**

47.77	Retail sale of watches and jewellery in specialised stores
47.78	Other retail sale of new goods in specialised stores
47.78/1	Retail sale in commercial art galleries

This subclass includes:

– activities of commercial art galleries

47.78/2   Retail sale by opticians

This subclass includes:

– specialised retail sale of spectacles, contact lenses, etc. by:
  - dispensing opticians
  - optometrists
  - opthalmic opticians, etc.

47.78/9   Other retail sale of new goods in specialised stores (other than by opticians or commercial art galleries), n.e.c.

This subclass includes:

– retail sale of photographic, optical and precision equipment
– retail sale of souvenirs, craftwork and religious articles
– retail sale of household fuel oil, bottled gas, coal and fuel wood
– retail sale of weapons and ammunition
– retail sale of stamps and coins
– retail sale of non-food products n.e.c.

47.79	Retail sale of second-hand goods in stores
47.79/1	Retail sale of antiques including antique books, in stores

This subclass includes:

– retail sale of antiques
– retail sale of antique books
– activities of auctioning houses (retail) (antiques)

This subclass excludes:

– *activities of Internet auctions and other non-store auctions (retail), see 47.91, 47.99*
– *activities of auctioning houses (retail) (second-hand goods), see 47.79/9*

47.79/9   Retail sale of second-hand goods (other than antiques and antique books) in stores

This subclass includes:

– retail sale of second-hand books
– retail sale of other second-hand goods
– activities of auctioning houses (retail) (second-hand goods)

This subclass excludes:

– *retail sale of second-hand motor vehicles, see 45.1*
– *activities of Internet auctions and other non-store auctions (retail), see 47.91, 47.99*
– *activities of pawn shops and pawnbrokers where the principal activity is lending money, see 64.92*
– *activities of auctioning houses (retail) (antiques), see 47.79/1*

47.8   Retail sale via stalls and markets

This group includes the retail sale of any kind of new or second-hand product in a usually movable stall either along a public road or at a fixed marketplace.

47.81   Retail sale via stalls and markets of food, beverages and tobacco products

This class excludes:

– *retail sale of prepared food for immediate consumption (mobile food vendors), see 56.10*

47.82   Retail sale via stalls and markets of textiles, clothing and footwear

47.89   Retail sale via stalls and markets of other goods

This class includes:

– retail sale of other goods via stalls or markets, such as:
  - carpets and rugs

- books
- games and toys
- household appliances and consumer electronics
- music and video recordings

47.9   **Retail trade not in stores, stalls or markets**

This group includes retail sale activities by mail order houses, over the Internet, through door-to-door sales, vending machines etc.

47.91   **Retail sale via mail order houses or via Internet**

This class includes retail sale activities via mail order houses or via Internet, i.e. retail sale activities where the buyer makes his choice on the basis of advertisements, catalogues, information provided on a website, models or any other means of advertising and places his order by mail, phone or over the Internet (usually through special means provided by a website). The products purchased can be either directly downloaded from the Internet or physically delivered to the customer.

This class includes:

- retail sale of any kind of product by mail order
- retail sale of any kind of product over the Internet

This class also includes:

- direct sale via television, radio and telephone
- Internet retail auctions

This class excludes:

- *retail sale of motor vehicles and motor vehicles parts and accessories over the Internet, see groups 45.1, 45.3*
- *retail sale of motorcycles and motorcycles parts and accessories over the Internet, see 45.40*

47.99   **Other retail sale not in stores, stalls or markets**

This class includes:

- retail sale of any kind of product in any way that is not included in previous classes:
  - by direct sales or door-to-door sales persons
  - through vending machines etc.
- direct selling of fuel (heating oil, firewood, etc.), delivered to the customers premises
- activities of non-store auctions (retail, except Internet)
- retail sale by (non-store) commission agents

Section H　　　　**Transportation and Storage**

This section includes the provision of passenger or freight transport, whether scheduled or not, by rail, pipeline, road, water or air and associated activities such as terminal and parking facilities, cargo handling, storage etc. Included in this section is the renting of transport equipment with driver or operator. Also included are postal and courier activities.

This section excludes:

– *major repair or alteration of transport equipment, except motor vehicles, see group 33.1*
– *construction, maintenance and repair of roads, railways, harbours, airfields, see division 42*
– *maintenance and repair of motor vehicles, see 45.20*
– *renting of transport equipment without driver or operator, see 77.1, 77.3*

49　　　　**Land transport and transport via pipelines**

This division includes the transport of passengers and freight via road and rail, as well as freight transport via pipelines.

49.1　　　　**Passenger rail transport, interurban**

49.10　　　　Passenger rail transport, interurban

This class includes:

– rail transportation of passengers using railway rolling stock on mainline networks, spread over an extensive geographic area
– passenger transport by interurban railways
– operation of sleeping cars or dining cars as an integrated operation of railway companies

This class excludes:

– *passenger transport by urban and suburban transit systems, see 49.31*
– *passenger terminal activities, see 52.21*
– *operation of railway infrastructure; related activities such as switching and shunting, see 52.21*
– *operation of sleeping cars or dining cars when operated by separate units, see 55.90, 56.10*

49.2　　　　**Freight rail transport**

49.20　　　　Freight rail transport

This class includes:

– freight transport on mainline rail networks as well as short line freight railways

This class excludes:

– *warehousing and storage, see 52.10*
– *freight terminal activities, see 52.21*
– *operation of railway infrastructure; related activities such as switching and shunting, see 52.21*
– *cargo handling, see 52.24*

49.3　　　　**Other passenger land transport**

This group includes all land-based passenger transport activities other than rail transport. However, rail transport as part of urban or suburban transport systems is included here.

49.31　　　　Urban and suburban passenger land transport

49.31/1　　　　Urban, suburban or metropolitan area passenger railway transportation by underground, metro and similar systems

This subclass includes:

– land transport of passengers by urban, suburban or metropolitan underground and elevated railways etc. The transport is carried out on scheduled routes normally following a fixed time schedule, entailing the picking up and setting down of passengers at normally fixed stops

This subclass also includes:

– town-to-airport or town-to-station transport by rail

This subclass excludes:

– *passenger transport by interurban railways, see 49.10*
– *tramways, see 49.31/9*

49.31/9　　　　Urban, suburban or metropolitan area passenger land transport other than railway transportation by underground, metro and similar systems

This subclass includes:

– land transport of passengers by urban, suburban or metropolitan transport systems. This may include different modes of land transport, such as by motor bus, tramway, streetcar, trolley bus. The transport is carried out on scheduled

171

routes normally following a fixed time schedule, entailing the picking up and setting down of passengers at normally fixed stops

This subclass also includes:

– town-to-airport or town-to-station transport by bus etc.
– operation of funicular railways, aerial cableways etc. if part of urban, suburban or metropolitan transit systems

This subclass excludes:

– *urban, suburban or metropolitan area passenger railway transportation by underground, metro and similar systems*

49.32   Taxi operation

This class also includes:

– other renting of private cars with driver

49.39   Other passenger land transport n.e.c.

This class includes:

– other passenger road transport:
    ■ scheduled long-distance bus services
    ■ charters, excursions and other occasional coach services
    ■ airport shuttles
– operation of teleferics, funiculars, ski and cable lifts if not part of urban, or suburban transit systems

This class also includes:

– operation of school buses and buses for transport of employees
– passenger transport by man- or animal-drawn vehicles

This class excludes:

– *ambulance transport, see 86.90*

49.4   Freight transport by road and removal services

This group includes all land-based freight transport activities other than rail transport.

49.41   Freight transport by road

This class includes:

– all freight transport operations by road:
    ■ logging haulage
    ■ stock haulage
    ■ refrigerated haulage
    ■ heavy haulage
    ■ bulk haulage, including haulage in tanker trucks including milk collection at farms
    ■ haulage of automobiles
    ■ transport of waste and waste materials, without collection or disposal

This class also includes:

– renting of trucks with driver
– freight transport by man- or animal-drawn vehicles

This class excludes:

– *log hauling within the forest, as part of logging operations, see 02.40*
– *distribution of water by trucks, see 36.00*
– *operation of terminal facilities for handling freight, see 52.21*
– *crating and packing activities for transport, see 52.29*
– *post and courier activities, see 53.10, 53.20*
– *waste transport as integrated part of waste collection activities, see 38.11, 38.12*

49.42   Removal services

This class includes:

– removal (relocation) services to businesses and households by road transport.

49.5   Transport via pipeline

49.50   Transport via pipeline

This class includes:

- transport of gases, liquids, water, slurry and other commodities via pipelines

This class also includes:

- operation of pump stations

This class excludes:

- *distribution of natural or manufactured gas, steam or water, see 35.22, 35.30, 36.00*
- *transport of liquids by trucks, see 49.41*

50   **Water transport**

This division includes the transport of passengers or freight over water, whether scheduled or not. Also included are the operation of towing or pushing boats, excursion, cruise or sightseeing boats, ferries, water taxis etc. Although the location is an indicator for the separation between sea and inland water transport, the deciding factor is the type of vessel used. Transport on sea-going vessels is classified in groups 50.1 and 50.2, while transport using other vessels is classified in groups 50.3 and 50.4.

This division excludes restaurant and bar activities on board ships (see 56.10, 56.30), if carried out by separate units.

50.1   **Sea and coastal passenger water transport**

This group includes the transport of passengers on vessels designed for operating on sea or coastal waters. Also included is the transport of passengers on great lakes etc. when similar types of vessels are used.

50.10   Sea and coastal passenger water transport

This class includes:

- transport of passengers over seas and coastal waters, whether scheduled or not:
    - operation of excursion, cruise or sightseeing boats
    - operation of ferries, water taxis etc.

This class also includes:

- renting of pleasure boats with crew for sea and coastal water transport (e.g. for fishing cruises)

This class excludes:

- *restaurant and bar activities on board ships, when provided by separate units, see 56.10, 56.30*
- *renting of pleasure boats and yachts without crew, see 77.21*
- *renting of commercial ships or boats without crew, see 77.34*
- *operation of "floating casinos", see 92.00*

50.2   **Sea and coastal freight water transport**

This group includes the transport of freight on vessels designed for operating on sea or coastal waters. Also included is the transport of freight on great lakes etc. when similar types of vessels are used.

50.20   Sea and coastal freight water transport

This class includes:

- transport of freight over seas and coastal waters, whether scheduled or not
- transport by towing or pushing of barges, oil rigs etc.

This class also includes:

- renting of vessels with crew for sea and coastal freight water transport

This class excludes:

- *storage of freight, see 52.10*
- *harbour operation and other auxiliary activities such as docking, pilotage, lighterage, vessel salvage, see 52.22*
- *cargo handling, see 52.24*
- *renting of commercial ships or boats without crew, see 77.34*

50.3   **Inland passenger water transport**

This group includes the transport of passengers on inland waters, involving vessels that are not suitable for sea transport.

50.30   Inland passenger water transport

This class includes:

- transport of passengers via rivers, canals, lakes and other inland waterways, including inside harbours and ports

This class also includes:

– renting of pleasure boats with crew for inland water transport

This class excludes:

– *renting of pleasure boats and yachts without crew, see 77.21*

50.4 **Inland freight water transport**

This group includes the transport of freight on inland waters, involving vessels that are not suitable for sea transport.

50.40 Inland freight water transport

This class includes:

– transport of freight via rivers, canals, lakes and other inland waterways, including inside harbours and ports

This class also includes:

– renting of vessels with crew for inland freight water transport

This class excludes:

– *cargo handling, see 52.24*
– *renting of commercial ships or boats without crew, see 77.34*

51 **Air transport**

This division includes the transport of passengers or freight by air or via space.

This division excludes:

– *crop spraying, see 01.61*
– *overhaul of aircraft or aircraft engines, see 33.16*
– *operation of airports, see 52.23*
– *aerial advertising (sky-writing), see 73.11*
– *aerial photography, see 74.20*

51.1 **Passenger air transport**

51.10 Passenger air transport

51.10/1 Scheduled passenger air transport

This subclass includes:

– transport of passengers by air over regular routes and on regular schedules

This subclass also includes:

– renting of air-transport equipment with operator for the purpose of scheduled passenger transportation

This subclass excludes:

– *renting of air transport equipment without operator, see 77.35*

51.10/2 Non-scheduled passenger air transport

This subclass includes:

– non-scheduled transport of passengers by air
– scenic and sightseeing flights
– regular charter flights for passengers
– general aviation activities, such as:
  - transport of passengers by aero clubs for instruction or pleasure

51.2 **Freight air transport and space transport**

51.21 Freight air transport

This class includes:

– transport freight by air over regular routes and on regular schedules
– non-scheduled transport of freight by air

This class also includes:

– renting of air transport equipment with operator for the purpose of freight transportation

51.22   Space transport

This class includes:

- launching of satellites and space vehicles
- space transport of freight and passengers

52   **Warehousing and support activities for transportation**

This division includes warehousing and support activities for transportation, such as operating of transport infrastructure (e.g. airports, harbours, tunnels, bridges, etc.), the activities of transport agencies and cargo handling.

52.1   **Warehousing and storage**

52.10   Warehousing and storage

52.10/1   Operation of warehousing and storage facilities for water transport activities of division 50

This subclass includes:

- operation of storage and warehouse facilities for all kind of goods:
  - operation of grain silos, general merchandise warehouses, refrigerated warehouses, storage tanks etc.

This subclass also includes:

- storage of goods in foreign trade zones
- blast freezing

This subclass excludes:

- *parking facilities for motor vehicles, see 52.21*
- *operation of self storage facilities, see 68.20*
- *rental of vacant space, see 68.20*

52.10/2   Operation of warehousing and storage facilities for air transport activities of division 51

This subclass includes:

- operation of storage and warehouse facilities for all kind of goods:
  - operation of grain silos, general merchandise warehouses, refrigerated warehouses, storage tanks etc.

This subclass also includes:

- storage of goods in foreign trade zones
- blast freezing

This subclass excludes:

- *parking facilities for motor vehicles, see 52.21*
- *operation of self storage facilities, see 68.20*
- *rental of vacant space, see 68.20*

52.10/3   Operation of warehousing and storage facilities for land transport activities of division 49

This subclass includes:

- operation of storage and warehouse facilities for all kind of goods:
  - operation of grain silos, general merchandise warehouses, refrigerated warehouses, storage tanks etc.

This subclass also includes:

- storage of goods in foreign trade zones
- blast freezing

This subclass excludes:

- *parking facilities for motor vehicles, see 52.21*
- *operation of self storage facilities, see 68.20*
- *rental of vacant space, see 68.20*

52.2   **Support activities for transportation**

This group includes activities supporting the transport of passengers or freight, such as operation of parts of the transport infrastructure or activities related to handling freight immediately before or after transport or between transport segments. The operation and maintenance of all transport facilities is included.

52.21   **Service activities incidental to land transportation**

52.21/1   Operation of rail freight terminals

This subclass excludes:

– *cargo handling, see 52.24/3*

52.21/2   Operation of rail passenger facilities at railway stations

This subclass excludes:

– *cargo handling, see 52.24/3*

52.21/3   Operation of bus and coach passenger facilities at bus and coach stations

This subclass excludes:

– *cargo handling, see 52.24/3*

52.21/9   Other service activities incidental to land transportation, n.e.c. (not including operation of rail freight terminals, passenger facilities at railway stations or passenger facilities at bus and coach stations)

This subclass includes:

– operation of roads, bridges, tunnels, car parks or garages, bicycle parks, winter storage of caravans
– towing and road side assistance
– switching and shunting
– other activities related to land transport of passengers, animals or freight:

This subclass also includes:

– liquefaction of gas for transportation purposes

This subclass excludes:

– *Operation of rail freight terminals, see 52.21/1*
– *Operation of rail passenger facilities at railway stations, see 52.21/2*
– *Operation of bus and coach passenger facilities at bus and coach stations, see 52.21/3*
– *cargo handling, see 52.24/3*

52.22   Service activities incidental to water transportation

This class includes:

– activities related to water transport of passengers, animals or freight:
  - operation of terminal facilities such as harbours and piers
  - operation of waterway locks etc.
  - navigation, pilotage and berthing activities
  - lighterage, salvage activities
  - lighthouse activities

This class excludes:

– *cargo handling, see 52.24*
– *operation of marinas, see 93.29*

52.23   Service activities incidental to air transportation

This class includes:

– activities related to air transport of passengers, animals or freight:
  - operation of terminal facilities such as airway terminals etc.
  - airport and air-traffic-control activities
  - ground service activities on airfields etc.

This class also includes:

– firefighting and fire-prevention services at airports

This class excludes:

– *cargo handling, see 52.24*
– *operation of flying schools, see 85.32, 85.53*

52.24   Cargo handling

52.24/1   Cargo handling for water transport activities of division 50

This subclass includes:

– loading and unloading of goods or passengers' luggage travelling via water transport

– stevedoring

This subclass excludes:

– *operation of terminal facilities, see 52.22*

52.24/2   Cargo handling for air transport activities of division 51

This subclass includes:

– loading and unloading of goods or passengers' luggage travelling via air transport

This subclass excludes:

– *operation of terminal facilities, see 52.23*

52.24/3   Cargo handling for land transport activities of division 49

This subclass includes:

– loading and unloading of goods or passengers' luggage travelling via rail transportation
– loading and unloading of freight railway cars

This subclass excludes:

– *operation of terminal facilities, see 52.21*

52.29   Other transportation support activities

This class includes:

– forwarding of freight
– arranging or organising of transport operations by rail, road, sea or air
– organisation of group and individual consignments (including pickup and delivery of goods and grouping of consignments)
– issue and procurement of transport documents and waybills
– activities of customs agents
– activities of sea-freight forwarders and air-cargo agents
– brokerage for ship and aircraft space
– goods-handling operations, e.g. temporary crating for the sole purpose of protecting the goods during transit, uncrating, sampling, weighing of goods

This class excludes:

– *courier activities, see 53.20*
– *provision of motor, marine, aviation and transport insurance, see 65.12*
– *activities of travel agencies, see 79.11*
– *activities of tour operators, see 79.12*
– *tourist assistance activities, see 79.90*

53   **Postal and courier activities**

This division includes postal and courier activities, such as pickup, transport and delivery of letters and parcels under various arrangements. Local delivery and messenger services are also included.

53.1   **Postal activities under universal service obligation**

53.10   Postal activities under universal service obligation

This class includes the activities of postal services operating under a universal service obligation by one or more designated universal service providers. The activities include use of the universal service infrastructure, including retail locations, sorting and processing facilities, and carrier routes to pickup and deliver the mail. The delivery can include letter-post, i.e. letters, postcards, printed papers (newspaper, periodicals, advertising items, etc.), small packets, goods or documents. Also included are other services necessary to support the universal service obligation.

This class includes:

– pickup, sorting, transport and delivery (domestic or international) of letter-post and (mail-type) parcels and packages by postal services operating under a universal service obligation. One or more modes of transport may be involved and the activity may be carried out with either self-owned (private) transport or via public transport.
– collection of letter-mail and parcels from public letter-boxes or from post offices

This class excludes:

– *postal giro, postal savings activities and money order activities, see 64.19*

53.2   **Other postal and courier activities**

53.20   Other postal and courier activities

53.20/1   **Licensed carriers**

This subclass includes:

– pickup, sorting, transport and delivery (domestic or international) of letter-post and (mail-type) parcels and packages by firms operating outside the scope of a universal service obligation. One or more modes of transport may be involved and the activity may be carried out with either self-owned (private) transport or via public transport.

This subclass also includes:

– home delivery services

This subclass excludes:

– *transport of freight, see (according to mode of transport) 49.20, 49.41, 50.20, 50.40, 51.21, 51.22*

53.20/2   **Unlicensed carriers**

This subclass includes:

– pickup, sorting, transport and delivery (domestic or international) of (mail-type) parcels and packages by firms operating outside the scope of a universal service obligation. One or more modes of transport may be involved and the activity may be carried out with either self-owned (private) transport or via public transport.

This subclass also includes:

– home delivery services

This subclass excludes:

– *transport of freight, see (according to mode of transport) 49.20, 49.41, 50.20, 50.40, 51.21, 51.22*

Section I   **Accommodation and Food Service Activities**

This section includes the provision of short-stay accommodation for visitors and other travellers and the provision of complete meals and drinks fit for immediate consumption. The amount and type of supplementary services provided within this section can vary widely.

This section excludes the provision of long-term accommodation as primary residences, which is classified in real estate activities (section L). Also excluded is the preparation of food or drinks that are either not fit for immediate consumption or that are sold through independent distribution channels, i.e. through wholesale or retail trade activities. The preparation of these foods is classified in manufacturing (section C).

55   **Accommodation**

This division includes the provision of short-stay accommodation for visitors and other travellers. Also included is the provision of longer term accommodation for students, workers and similar individuals. Some units may provide only accommodation while others provide a combination of accommodation, meals and/or recreational facilities.

This division excludes activities related to the provision of long-term primary residences in facilities such as apartments typically leased on a monthly or annual basis classified in Real Estate (section L).

55.1   **Hotels and similar accommodation**

55.10   Hotels and similar accommodation

This class includes the provision of accommodation, typically on a daily or weekly basis, principally for short stays by visitors. This includes the provision of furnished accommodation in guest rooms and suites. Services include daily cleaning and bed-making. A range of additional services may be provided such as food and beverage services, parking, laundry services, swimming pools and exercise rooms, recreational facilities as well as conference and convention facilities.

This class includes accommodation provided by:

– hotels
– resort hotels
– suite/apartment hotels
– motels

This class excludes:

– *provision of homes and furnished or unfurnished flats or apartments for more permanent use, typically on a monthly or annual basis, see division 68*

55.2   **Holiday and other short-stay accommodation**

55.20   Holiday and other short-stay accommodation

This class includes the provision of accommodation, typically on a daily or weekly basis, principally for short stays by visitors, in self-contained space consisting of complete furnished rooms or areas for living/dining and sleeping, with cooking facilities or fully equipped kitchens. This may take the form of apartments or flats in small free-standing multi-storey buildings or clusters of buildings, or single storey bungalows, chalets, cottages and cabins. Very minimal complementary services, if any, are provided.

This class includes accommodation provided by:

– childrens and other holiday homes
– visitor flats and bungalows
– cottages and cabins without housekeeping services
– youth hostels
– mountain refuges

This class excludes:

– *provision of furnished short-stay accommodation with daily cleaning, bed-making, food and beverage services, see 55.10*
– *provision of homes and furnished or unfurnished flats or apartments for more permanent use, typically on a monthly or annual basis, see division 68*

55.20/1   Holiday centres and villages

This subclass includes:

– provision of holiday and other collective accommodation in holiday centres and holiday villages

55.20/2   Youth hostels

This subclass includes:

– mountain refuges

This subclass excludes:

– *protective shelters or plain bivouac facilities for placing tents and/or sleeping bags, see 55.30*

55.20/9   **Other holiday and other short-stay accommodation (not including holiday centres and villages or youth hostels) n.e.c.**

This subclass includes:

– provision of holiday and other collective accommodation other than that provided in holiday centres and holiday villages or in youth hostels

55.3   **Camping grounds, recreational vehicle parks and trailer parks**

55.30   Camping grounds, recreational vehicle parks and trailer parks

This class includes:

– provision of accommodation in campgrounds, trailer parks, recreational camps and fishing and hunting camps for short stay visitors.
– provision of space and facilities for recreational vehicles

This class also includes accommodation provided by:

– protective shelters or plain bivouac facilities for placing tents and/or sleeping bags

This class excludes:

– *mountain refuges, cabins and hostels, see 55.20*

55.9   **Other accommodation**

55.90   Other accommodation

This class includes the provision temporary or longer-term accommodation in single or shared rooms or dormitories for students, migrant (seasonal) workers and other individuals.

This class includes:

– student residences
– school dormitories
– workers hostels
– rooming and boarding houses
– railway sleeping cars

56   **Food and beverage service activities**

This division includes food and beverage serving activities providing complete meals or drinks fit for immediate consumption, whether in traditional restaurants, self-service or take-away restaurants, whether as permanent or temporary stands with or without seating. The fact that meals fit for immediate consumption are offered is the decisive factor rather than the kind of facility providing them.

This division excludes the production of meals not fit for immediate consumption or not planned to be consumed immediately or of prepared food which is not considered to be a meal (see divisions 10: manufacture of food products and 11: manufacture of beverages). Also excluded is the sale of not self-manufactured food that is not considered to be a meal or of meals that are not fit for immediate consumption (see section G: wholesale and retail trade).

56.1   **Restaurants and mobile food service activities**

56.10   Restaurants and mobile food service activities

56.10/1   Licensed restaurants

This subclass includes the provision of food services to customers, whether they are served while seated or serve themselves from a display of items. The meals provided are generally for consumption on the premises and alcoholic drinks to accompany the meal are available.

This subclass includes:

– restaurants
– cafeterias
– fast-food restaurants

The subclass also includes restaurant and bar activities connected to transportation, when carried out by separate units.

This subclass excludes:

– *concession operation of eating facilities, see 56.29*

56.10/2   Unlicensed restaurants and cafes

This subclass includes the provision of food services to customers, whether they are served while seated or serve themselves from a display of items, The meals provided are generally for consumption on the premises and only non-alcoholic drinks are served.

This subclass includes:

- restaurants
- cafeterias
- fast-food restaurants

The subclass also includes restaurant and bar activities connected to transportation, when carried out by separate units.

This subclass excludes:

- *concession operation of eating facilities, see 56.29*

56.10/3　Take away food shops and mobile food stands

This subclass includes the provision of food services to customers to take away or to have delivered. This includes the preparation and serving of meals for immediate consumption from motorised vehicles or non-motorised carts.

This subclass includes:

- take-out eating places
- ice cream truck vendors
- mobile food carts
- food preparation in market stalls

This subclass excludes:

- *retail sale of food through vending machines, see 47.99*
- *concession operation of eating facilities, see 56.29*

56.2　Event catering and other food service activities

This group includes catering activities for individual events or for a specified period of time and the operation of food concessions, such as at sports or similar facilities.

56.21　Event catering activities

This class includes the provision of food services based on contractual arrangements with the customer, at the location specified by the customer, for a specific event.

This class excludes:

- *manufacture of perishable food items for resale, see 10.89*
- *retail sale of perishable food items, see division 47*

56.29　Other food service activities

This class includes industrial catering, i.e. the provision of food services based on contractual arrangements with the customer, for a specific period of time. Also included is the operation of food concessions at sports and similar facilities. The food is usually prepared in a central unit.

This class includes:

- activities of food service contractors (e.g. for transportation companies)
- operation of food concessions at sports and similar facilities
- operation of canteens or cafeterias (e.g. for factories, offices, hospitals or schools) on a concession basis

This class excludes:

- *manufacture of perishable food items for resale, see 10.89*
- *retail sale of perishable food items, see division 47*

56.3　Beverage serving activities

This group includes the preparation and serving of beverages for immediate consumption on the premises.

56.30　Beverage serving activities
56.30/1　Licensed clubs

This subclass includes the preparation and serving of beverages for immediate consumption on the premises by:

- nightclubs
- social clubs

This subclass excludes:

- *reselling packaged/prepared beverages, see 47*
- *retail sale of beverages through vending machines, see 47.99*

56.30/2   Public houses and bars

This subclass includes the preparation and serving of beverages for immediate consumption on the premises by:

– bars
– taverns
– cocktail lounges
– discotheques licensed to sell alcohol (with beverage serving predominant)
– beer parlours

This subclass excludes:

– *reselling packaged/prepared beverages, see 47*
– *retail sale of beverages through vending machines, see 47.99*
– *operation of discotheques and dance floors without beverage serving, see 93.29*

Section J   **Information and Communication**

This section includes the production and distribution of information and cultural products, the provision of the means to transmit or distribute these products, as well as data or communications, information technology activities and the processing of data and other information service activities.

The main components of this section are publishing activities (division 58), including software publishing, motion picture and sound recording activities (division 59), radio and TV broadcasting and programming activities (division 60), telecommunications activities (division 61), information technology activities (division 62) and other information service activities (division 63).

Publishing includes the acquisition of copyrights for content (information products) and making this content available to the general public by engaging in (or arranging for) the reproduction and distribution of this content in various forms. All the feasible forms of publishing (in print, electronic or audio form, on the internet, as multimedia products such as CD-ROM reference books etc.) are included in this section.

Activities related to production and distribution of TV programming span divisions 59, 60 and 61, reflecting different stages in this process. Individual components, such as movies, television series etc. are produced by activities in division 59, while the creation of a complete television channel programme, from components produced in division 59 or other components (such as live news programming) is included in division 60. Division 60 also includes the broadcasting of this programme by the producer. The distribution of the complete television programme by third parties, i.e. without any alteration of the content, is included in division 61. This distribution in division 61 can be done through broadcasting, satellite or cable systems.

58   **Publishing activities**

This division includes the publishing of books, brochures, leaflets, dictionaries, encyclopaedias, atlases, maps and charts; publishing of newspapers, journals and periodicals; directory and mailing list and other publishing, as well as software publishing.

Publishing includes the acquisition of copyrights to content (information products) and making this content available to the general public by engaging in (or arranging for) the reproduction and distribution of this content in various forms. All the feasible forms of publishing (in print, electronic or audio form, on the internet, as multimedia products such as CD-ROM reference books etc.), except publishing of motion pictures, are included in this division.

This division excludes the publishing of motion pictures, video tapes and movies on DVD or similar media (division 59) and the production of master copies for records or audio material (division 59). Also excluded is printing (see 18.11, 18.12) and the mass reproduction of recorded media (see 18.20).

58.1   **Publishing of books, periodicals and other publishing activities**

This group includes activities of publishing books, newspapers, magazines and other periodicals, directories and mailing lists, and other works such as photos, engravings, postcards, timetables, forms, posters and reproductions of works of art. These works are characterised by the intellectual creativity required in their development and are usually protected by copyright.

58.11   Book publishing

This class includes the activities of publishing of books in print, electronic (CD, electronic displays etc.) or audio form or on the Internet.

Included are:

– publishing of books, brochures, leaflets and similar publications, including publishing of dictionaries and encyclopaedias
– publishing of atlases, maps and charts
– publishing of audio books
– publishing of encyclopaedias etc. on CD-ROM

This class excludes:

– *production of globes, see 32.99*
– *publishing of advertising material, see 58.19*
– *publishing of music and sheet books, see 59.20*
– *activities of independent authors, see 90.03*

58.12   Publishing of directories and mailing lists

This class includes the publishing of lists of facts/information (databases) that are protected in their form, but not in their content. These lists can be published in printed or electronic form.

This class includes:

– publishing of mailing lists
– publishing of telephone books
– publishing of other directories and compilations, such as case law, pharmaceutical compendia etc.

58.13   Publishing of newspapers

This class includes the publishing of newspapers, including advertising newspapers, appearing at least four times a week. Publishing can be in print or electronic form, including on the Internet.

This class excludes:

– *news agency activities, see 63.91*

58.14   Publishing of journals and periodicals

58.14/1   Publishing of learned journals

This subclass includes the activities of publishing journals reporting or discussing the results of scholarly research and intended for an academic and research-based readership; they are normally peer-reviewed and the primary source of revenue is not the sale of advertisement space, but subscription. They appear less than four times a week and can be published in print or electronic form, including on the Internet.

58.14/2   Publishing of consumer, business and professional journals and periodicals

This subclass includes the activities of publishing journals and periodicals providing entertainment or lifestyle information to individuals, information within businesses or across the business community or information for trades or professions. It also includes customer magazines produced for businesses and distributed free, newsletters and the publishing of radio and television schedules. These publications appear less than four times a week and can be published in print or electronic form, including on the Internet.

58.19   Other publishing activities

This class includes:

– publishing (including on-line) of:
  - catalogues
  - photos, engravings and postcards
  - greeting cards
  - forms
  - posters, reproduction of works of art
  - advertising material
  - other printed matter
– on-line publishing of statistics and other information

This class excludes:

– *publishing of advertising newspapers, see 58.13*
– *on-line provision of software (application hosting and application service provisioning), see 63.11*

58.2   **Software publishing**

58.21   Publishing of computer games

This class includes:

– publishing of computer games for all platforms

58.29   Other software publishing

This class includes:

– publishing of ready-made (non-customised) software, including translation or adaptation of non-customised software for a particular market on own account:
  - operating systems
  - business and other applications

This class excludes:

– *reproduction of software, see 18.20*
– *retail sale of non-customised software, see 47.41*
– *production of software not associated with publishing, including translation or adaptation of non-customised software for a particular market on a fee or contract basis, see 62.01*
– *on-line provision of software (application hosting and application service provisioning), see 63.11*

59   **Motion picture, video and television programme production, sound recording and music publishing activities**

This division includes production of theatrical and non-theatrical motion pictures whether on film, video tape or disc for direct projection in theatres or for broadcasting on television; supporting activities such as film editing, cutting, dubbing etc.; distribution of motion pictures and other film productions to other industries; as well as motion picture or other film productions projection. Buying and selling of motion picture or other film productions distribution rights is also included.

This division also includes sound recording activities, i.e. production of original sound master recordings, releasing, promoting and distributing them, publishing of music as well as sound recording service activities in a studio or elsewhere.

59.1　　　　Motion picture, video and television programme activities

This group includes production of theatrical and non-theatrical motion pictures whether on film, video tape, DVD or other media, including digital distribution, for direct projection in theatres or for broadcasting on television; supporting activities such as film editing, cutting, dubbing etc.; distribution of motion pictures or other film productions (video tapes, DVDs, etc) to other industries; as well as their projection. Buying and selling of motion picture or any other film production distribution rights is also included.

59.11　　　　Motion picture, video and television programme production activities

59.11/1　　　　Motion picture production activities

This subclass includes:

– production of motion pictures

This subclass excludes:

– *film duplicating (except reproduction of motion picture film for theatrical distribution) from master copies, see 18.20*
– *post-production activities, see 59.12*
– *sound recording and recording of books on tape, see 59.20*
– *film processing other than for the motion picture industry, see 74.20*
– *activities of personal theatrical or artistic agents or agencies, see 74.90*
– *activities of own account actors, cartoonists, directors, stage designers and technical specialists, see 90.0*

59.11/2　　　　Video production activities

This subclass includes:

– production of videos

This subclass excludes:

– *audio and video tape, CD or DVD reproduction from master copies, see 18.20*
– *wholesale of recorded video tapes, CD-s, DVD-s, see 46.43*
– *wholesale of blank video tapes, CD-s, see 46.52*
– *retail trade of video tapes, CD-s, DVD-s, see 47.63*
– *renting of video tapes, DVD-s to the general public, see 77.22*
– *post-production activities, see 59.12*
– *sound recording and recording of books on tape, see 59.20*
– *activities of personal theatrical or artistic agents or agencies, see 74.90*
– *activities of own account actors, cartoonists, directors, stage designers and technical specialists, see 90.0*

59.11/3　　　　Television programme production activities

This subclass includes:

– production of television programmes (televisions series, documentaries etc.), or television advertisements

This subclass excludes:

– *post-production activities, see 59.12*
– *sound recording and recording of books on tape, see 59.20*
– *television broadcasting, see 60.2*
– *creating a complete television channel programme, see 60.2*
– *activities of personal theatrical or artistic agents or agencies, see 74.90*
– *activities of own account actors, cartoonists, directors, stage designers and technical specialists, see 90.0*
– *real-time (i.e. simultaneous) closed captioning of live television performances of meetings, conferences, etc., see 82.99*

59.12　　　　Motion picture, video and television programme post-production activities

This class includes post-production activities such as editing, film/tape transfers, titling, subtitling, credits, closed captioning, computer-produced graphics, animation and special effects, developing and processing motion picture film, as well as activities of motion picture film laboratories and activities of special laboratories for animated films.

This class also includes:

– the activities of stock footage film libraries, etc.

This class excludes:

– *film duplicating (except reproduction of motion picture film for theatrical distribution) as well as audio and video tape, CD or DVD reproduction from master copies, see 18.20*

- *wholesale of recorded video tapes, CD-s, DVD-s, see 46.43*
- *wholesale of blank video tapes, CD-s, see 46.52*
- *retail trade of video tapes, CD-s, DVD-s, see 47.63*
- *film processing other than for the motion picture industry, see 74.20*
- *renting of video tapes, DVD-s to the general public, see 77.22*
- *activities of own account actors, cartoonists, directors, stage designers and technical specialists, see 90.0*

59.13   **Motion picture, video and television programme distribution activities**

59.13/1   **Motion picture distribution activities**

This subclass includes:

- distributing film to motion picture theatres, television networks and stations, and exhibitors.

This subclass also includes:

- acquiring film distribution rights

This subclass excludes:

- *film duplicating from master copies, see 18.20*

59.13/2   **Video distribution activities**

This subclass includes:

- distributing video tapes, DVD-s and similar productions to motion picture theatres, television networks and stations, and exhibitors

This subclass also includes:

- acquiring video tape and DVD distribution rights

This subclass excludes:

- *audio and video tape, CD or DVD reproduction from master copies, see 18.20*
- *wholesale of recorded video tapes and DVDs, see 46.43*
- *retail sale of recorded video tapes and DVDs, see 47.63*

59.13/3   **Television programme distribution activities**

This subclass includes:

- distributing television programmes to television networks and stations, and exhibitors

This subclass also includes:

- acquiring television distribution rights

59.14   **Motion picture projection activities**

This class includes:

- activities of motion picture or video tape projection in cinemas, in the open air or in other projection facilities
- activities of cine-clubs

59.2   **Sound recording and music publishing activities**

59.20   Sound recording and music publishing activities

This class includes the activities of production of original (sound) master recordings, such as tapes, CDs; releasing, promoting and distributing sound recordings to wholesalers, retailers or directly to the public. These activities might be integrated or not with the production of master recordings in the same unit. If not, the unit exercising these activities has to obtain the reproduction and distribution rights to master recordings. This class also includes sound recording service activities in a studio or elsewhere, including the production of taped (i.e. non-live) radio programming.

This class also includes the activities of music publishing, i.e. activities of acquiring and registering copyrights for musical compositions, promoting, authorising and using these compositions in recordings, radio, television, motion pictures, live performances, print and other media. Units engaged in these activities may own the copyright or act as administrator of the music copyrights on behalf of the copyright owners. Publishing of music and sheet books is included here.

60   **Programming and broadcasting activities**

This division includes the activities of creating content or acquiring the right to distribute content and subsequently broadcasting that content, such as radio, television and data programmes of entertainment, news, talk, and the like. Also included is data broadcasting, typically integrated with radio or TV broadcasting. The broadcasting can be performed using different technologies, over-the-air, via satellite, via a cable network or via Internet.

This division also includes the production of programmes that are typically narrowcast in nature (limited format, such as news, sports, education, and youth-oriented programming) on a subscription or fee basis, to a third party, for subsequent broadcasting to the public.

This division excludes the distribution of cable and other subscription programming (see division 61).

60.1 Radio broadcasting

60.10 Radio broadcasting

This class includes:

– activities of broadcasting audio signals through radio broadcasting studios and facilities for the transmission of aural programming to the public, to affiliates or to subscribers

This class also includes:

– activities of radio networks, i.e. assembling and transmitting aural programming to affiliates or subscribers via over-the-air broadcasts, cable or satellite
– radio broadcasting activities over the Internet (Internet radio stations)
– data broadcasting integrated with radio broadcasting.

This class excludes:

– *the production of taped radio programming, see 59.20*

60.2 Television programming and broadcasting activities

60.20 Television programming and broadcasting activities

This class includes the activities of creating a complete television channel programme, from purchased programme components (e.g. movies, documentaries etc.), self produced programme components (e.g. local news, live reports) or a combination thereof.

This complete television programme can be either broadcast by the producing unit or produced for transmission by a third party distributor, such as cable companies or satellite television providers.

The programming may be of a general or specialised nature (e.g. limited formats such as news, sports, education or youth oriented programming). This class includes programming that is made freely available to users, as well as programming that is available only on a subscription basis. The programming of video-on-demand channels is also included here.

This class also includes data broadcasting integrated with television broadcasting.

This class excludes:

– *the production of television programme elements (movies, documentaries, talk shows, commercials etc.) not associated with broadcasting, see 59.11*
– *the assembly of a package of channels and distribution of that package, without programming, see division 61*

61 Telecommunications

This division includes the activities of providing telecommunications and related service activities, that is transmitting voice, data, text, sound and video. The transmission facilities that carry out these activities may be based on a single technology or a combination of technologies. The common feature of the activities classified in this division is the transmission of content, without being involved in its creation. The breakdown in this division is based on the type of infrastructure operated.

In the case of transmission of television signals this may include the bundling of complete programming channels (produced in division 60) in to programme packages for distribution.

61.1 Wired telecommunications activities

61.10 Wired telecommunications activities

This class includes:

– operating, maintaining or providing access to facilities for the transmission of voice, data, text, sound and video using a wired telecommunications infrastructure, including:
  ▪ operating and maintaining switching and transmission facilities to provide point-to-point communications via landlines, microwave or a combination of landlines and satellite linkups
  ▪ operating of cable distribution systems (e.g. for distribution of data and television signals)
  ▪ furnishing telegraph and other non-vocal communications using own facilities

The transmission facilities that carry out these activities, may be based on a single technology or a combination of technologies.

This class also includes:

– purchasing access and network capacity from owners and operators of networks and providing telecommunications services using this capacity to businesses and households
– provision of Internet access by the operator of the wired infrastructure

This class excludes:

- *telecommunications resellers, see 61.90*

61.2   **Wireless telecommunications activities**

61.20   Wireless telecommunications activities

This class includes:

- operating, maintaining or providing access to facilities for the transmission of voice, data, text, sound, and video using a wireless telecommunications infrastructure
- maintaining and operating paging as well as cellular and other wireless telecommunications networks

The transmission facilities provide omni-directional transmission via airwaves and may be based on a single technology or a combination of technologies.

This class also includes:

- purchasing access and network capacity from owners and operators of networks and providing wireless telecommunications services (except satellite) using this capacity to businesses and households
- provision of Internet access by the operator of the wireless infrastructure

This class excludes:

- *telecommunications resellers, see 61.90*

61.3   **Satellite telecommunications activities**

61.30   Satellite telecommunications activities

This class includes:

- operating, maintaining or providing access to facilities for the transmission of voice, data, text, sound and video using a satellite telecommunications infrastructure
- delivery of visual, aural or textual programming received from cable networks, local television stations or radio networks to consumers via direct-to-home satellite systems. (The units classified here do not generally originate programming material.)

This class also includes:

- provision of Internet access by the operator of the satellite infrastructure

This class excludes:

- *telecommunications resellers, see 61.90*

61.9   **Other telecommunications activities**

61.90   Other telecommunications activities

This class includes:

- provision of specialised telecommunications applications, such as satellite tracking, communications telemetry, and radar station operations
- operation of satellite terminal stations and associated facilities operationally connected with one or more terrestrial communications systems and capable of transmitting telecommunications to or receiving telecommunications from satellite systems
- provision of Internet access over networks between the client and the ISP not owned or controlled by the ISP, such as dial-up Internet access etc.
- provision of telephone and Internet access in facilities open to the public
- provision of telecommunications services over existing telecom connections:
  - VOIP (Voice Over Internet Protocol) provision
- telecommunications resellers (i.e. purchasing and reselling network capacity without providing additional services)

This class excludes:

- *provision of Internet access by operators of telecommunications infrastructure, see 61.10, 61.20, 61.30*

62   **Computer programming, consultancy and related activities**

This division includes the following activities of providing expertise in the field of information technologies: writing, modifying, testing and supporting software; planning and designing computer systems that integrate computer hardware, software and communication technologies; on-site management and operation of clients' computer systems and/or data processing facilities; and other professional and technical computer-related activities.

62.0   **Computer programming, consultancy and related activities**

62.01　　Computer programming activities

This class includes the writing, modifying, testing and supporting of software.

This class includes:

- designing the structure and content of, and/or writing the computer code necessary to create and implement:
    - systems software (including updates and patches)
    - software applications (including updates and patches)
    - databases
    - web pages
- customising of software, i.e. modifying and configuring an existing application so that it is functional within the clients' information system environment

This class excludes:

- *publishing packaged software, see 58.29*
- *translation or adaptation of non-customised software for a particular market on own account, see 58.29*
- *planning and designing computer systems that integrate computer hardware, software and communication technologies, even though providing software might be an integral part, see 62.02*

62.01/1　　Ready-made interactive leisure and entertainment software development

This subclass includes the development, production, supply and documentation of ready-made interactive leisure and entertainment software, such as games software, designed for publication by a different enterprise. A key component part of the software is audiovisual content with which the user interacts. The software can be published across any format, such as games consoles, the internet and mobile phones.

62.01/2　　Business and domestic software development

This subclass excludes:

- *Ready-made interactive leisure and entertainment software development, see 62.01/1*

62.02　　Computer consultancy activities

This class includes the planning and designing of computer systems which integrate computer hardware, software and communication technologies. Services may include related users training.

This class excludes:

- *sale of computer hardware or software, see 46.51, 47.41*
- *installation of mainframe and similar computers, see 33.20*
- *installation (setting-up) of personal computers, see 62.09*
- *installation of software, computer disaster recovery, see 62.09*

62.03　　Computer facilities management activities

This class includes the provision of on-site management and operation of clients' computer systems and/or data processing facilities, as well as related support services.

62.09　　Other information technology and computed service activities

This class includes other information technology and computer related activities not elsewhere classified, such as:

- computer disaster recovery services
- installation (setting-up) of personal computers
- software installation services

This class excludes:

- *installation of mainframe and similar computers, see 33.20*
- *computer programming, see 62.01*
- *computer consultancy, see 62.02*
- *computer facilities management, see 62.03*
- *data processing and hosting, see 63.11*

63　　**Information service activities**

This division includes the activities of web search portals, data processing and hosting activities, as well as other activities that primarily supply information.

63.1　　**Data processing, hosting and related activities; web portals**

This group includes the provision of infrastructure for hosting, data processing services and related activities, as well as the provision of search facilities and other portals for the Internet.

63.11	**Data processing, hosting and related activities**

This class includes:

– provision of infrastructure for hosting, data processing services and related activities
– specialised hosting activities such as:
  ▪ Web hosting
  ▪ streaming services
  ▪ application hosting
– application service provisioning
– general time-share provision of mainframe facilities to clients
– data processing activities:
  ▪ complete processing of data supplied by clients
  ▪ generation of specialised reports from data supplied by clients
– provision of data entry services

This class excludes:

– *activities where the supplier uses the computers only as a tool are classified according to the nature of the services rendered*

63.12	**Web portals**

This class includes:

– the operation of websites that use a search engine to generate and maintain extensive databases of Internet addresses and content in an easily searchable format
– operation of other websites that act as portals to the Internet, such as media sites providing periodically updated content

This class excludes:

– *publishing of books, newspapers, journals etc. via Internet, see division 58*
– *broadcasting via Internet, see division 60*

63.9	**Other information service activities**

This group includes the activities of news agencies and all other remaining information service activities.

This group excludes:

– *activities of libraries and archives, see 91.01*

63.91	**News agency activities**

This class includes:

– news syndicate and news agency activities furnishing news, pictures and features to the media

This class excludes:

– *activities of independent photojournalists, see 74.20*
– *activities of independent journalists, see 90.03*

63.99	**Other information service activities n.e.c.**

This class includes other information service activities not elsewhere classified such as:

– computer-based telephone information services
– information search services on a contract or fee basis
– news clipping services, press clipping services, etc.

This class excludes:

– *activities of call centres, see 82.20*

Section K   **Financial and Insurance Activities**

This section includes financial service activities, including insurance, reinsurance and pension funding activities, and activities to support financial services.

This section also includes the activities of holding assets, such as activities of holding companies and the activities of trusts, funds and similar financial entities.

64   **Financial service activities, except insurance and pension funding**

This division includes the activities of obtaining and redistributing funds other than for the purpose of insurance or pension funding or compulsory social security.

Note: National institutional arrangements are likely to play a significant role in determining the classification within this division.

64.1   **Monetary intermediation**

This group includes the obtaining of funds in the form of transferable deposits, i.e. funds that are fixed in money terms, obtained on a day-to-day basis and, apart from central banking, obtained from non-financial sources.

64.11   Central banking

This class includes:

- issuing and managing the country's currency
- monitoring and control of the money supply
- taking deposits that are used for clearance between financial institutions
- supervising banking operations
- holding the country's international reserves
- acting as banker to the government

The activities of central banks will vary for institutional reasons.

64.19   Other monetary intermediation

This class includes the receiving of deposits and/or close substitutes for deposits and extending of credit or lending funds. The granting of credit can take a variety of forms, such as loans, mortgages, credit cards etc. These activities are generally carried out by monetary institutions other than central banks, such as:

- banks
- building societies
- savings banks
- credit unions

This class also includes:

- postal giro and postal savings bank activities
- credit granting for house purchase by specialised deposit-taking institutions
- money order activities

This class excludes:

- *credit granting for house purchase by specialised non-depository institutions, see 64.92*
- *credit card transaction processing and settlement activities, see 66.19*

64.19/1   Banks

This subclass includes only:

- monetary intermediation by those institutions other than the Bank of England, authorised by the Financial Services Authority under the Banking Act of 1987 to accept deposits in the UK
- monetary intermediation by those European authorised institutions which are entitled to accept deposits through a branch in the UK on the basis of their home state authorisation
- monetary intermediation of the National Savings Bank
- monetary intermediation of the Savings Certificate Office

64.19/2   Building societies

This subclass includes:

- monetary intermediation by those institutions authorised by the Financial Services Authority under the Building Societies Acts of 1986 and 1997

This subclass excludes:

- *activities of specialist mortgage finance companies see 64.92/2*
- *activities of housing associations see 41.10, 68.20/1*

64.2   **Activities of holding companies**

64.20   Activities of holding companies

This class includes the activities of holding companies, i.e. units that hold the assets (owning controlling- levels of equity) of a group of subsidiary corporations and whose principal activity is owning the group. The holding companies in this class do not provide any other service to the businesses in which the equity is held, i.e. they do not administer or manage other units.

This class excludes:

– active management of companies and enterprises, strategic planning and decision making of the company, see 70.10

64.20/1   Activities of agricultural holding companies

64.20/2   Activities of production holding companies

64.20/3   Activities of construction holding companies

64.20/4   Activities of distribution holding companies

64.20/5   Activities of financial services holding companies

64.20/9   Activities of other holding companies (not including agricultural, production, construction, distribution and financial services holding companies) n.e.c.

64.3   **Trusts, funds and similar financial entities**

64.30   Trusts, funds and similar financial entities

This class includes legal entities organised to pool securities or other financial assets, without managing, on behalf of shareholders or beneficiaries. The portfolios are customised to achieve specific investment characteristics, such as diversification, risk, rate of return, and price volatility. These entities earn interest, dividends, and other property income, but have little or no employment and no revenue from the sale of services.

This class includes:

– open-end investment funds
– closed-end investment funds
– trusts, estates or agency accounts, administered on behalf of the beneficiaries under the terms of a trust agreement, will or agency agreement
– unit investment trust funds

This class excludes:

– funds and trusts that earn revenue from the sale of goods or services, see SIC class according to their principal activity
– activities of holding companies, see 64.20
– pension funding, see 65.30
– management of funds, see 66.30

64.30/1   Activities of investment trusts

This subclass includes only:

– those investment trust companies recognised as such by HM Revenue & Customs for tax purposes

64.30/2   Activities of unit trusts

This subclass includes only:

– activities of unit trusts authorised by the Financial Services Authority under the terms of the Financial Services Act 1986 and 2000
– activities of unauthorised unit trusts such as "in-house" trusts (i.e. funds run on unit trust lines by, e.g., stockbrokers and merchant banks which are designed for their own clients)

64.30/3   Activities of venture and development capital companies

This subclass includes:

– activities of venture and development capital companies and funds

This subclass excludes:

– activities of venture capital investment trusts, see 64.30/1
– activities of authorised and unauthorised unit trusts, see 64.30/2
– activities of open-ended investment companies, see 64.30/4

64.30/4   Activities of open-ended investment companies

This subclass includes only:

– open-ended investment companies authorised by the Financial Services Authority under the Open-Ended Investment Companies (Companies with Variable Capital) Regulations

64.30/5　　Activities of property unit trusts

Property Unit Trusts have an open-ended share capital structure and hold, manage and maintain real estate portfolios (which could consist of a single property), for investment purposes (for renting to tenants), but could also be used to hold property for development or under construction. Units in these vehicles are typically traded through the fund manager and are not traded on a stock exchange.

64.30/6　　Activities of real estate investment trusts

Real Estate Investment Trust Vehicles are companies with closed-ended share capital that mainly hold, manage and maintain real estate portfolios, for investment purposes (for renting to tenants) and whose shares are listed and traded on a stock exchange.

64.9　　**Other financial service activities, except insurance and pension funding**

This group includes financial service activities other than those conducted by monetary institutions.

This group excludes:

– *insurance and pension funding, see division 65*

64.91　　Financial leasing

This class includes:

– leasing where the term approximately covers the expected life of the asset and the lessee acquires substantially all the benefits of its use and takes all the risks associated with its ownership. The ownership of the asset may or may not eventually be transferred. Such leases cover all or virtually all costs including interest.

This class excludes:

– *operational leasing, see division 77, according to type of goods leased*

64.92　　Other credit granting

This class includes:

– financial service activities primarily concerned with making loans by institutions not involved in monetary intermediation, where the granting of credit can take a variety of forms, such as loans, mortgages, credit cards etc. providing the following types of services:
  - granting of consumer credit
  - international trade financing
  - provision of long-term finance to industry by industrial banks
  - money lending outside the banking system
  - credit granting for house purchase by specialised non-depository institutions
  - pawnshops and pawnbrokers where the principal activity is lending money

This class excludes:

– *credit granting for house purchase by specialised institutions that also take deposits, see 64.19*
– *operational leasing, see division 77, according to type of goods leased*
– *grant giving activities by membership organisations, see 94.99*
– *pawnshops and pawnbrokers where the principal activity is dealing in second hand goods, see 47.79/2*

64.92/1　　Credit granting by non-deposit taking finance houses and other specialist consumer credit grantors

This subclass includes only:

– activities of non-deposit taking finance houses
– activities of hire purchase and loan companies not in the UK banking sector
– activities of check traders
– activities of money lenders
– pawnbroking where the primary activity is in lending money
– activities of building societies' personal finance subsidiaries
– other consumer credit granting where the main business is the direct financing (other than finance leasing) of instalment credit sales mainly to persons, together with farm, industrial and building plant equipment to unincorporated businesses

This subclass excludes:

– *activities of pawnshops where the primary activity is dealing in second-hand goods, see 47.79/2*
– *financial leasing, see 64.91*
– *operating leasing, see 77*

64.92/2   Activities of mortgage finance companies

This subclass includes only:

– activities of specialist mortgage finance companies other than banks and building societies

64.92/9   Other credit granting (not including credit granting by non-deposit taking finance houses and other specialist consumer credit grantors and activities of mortgage finance companies) n.e.c.

This subclass includes:

– activities of other institutions not in the UK banking sector whose main activity is to extend credit abroad
– activities of other special finance agencies and export credit companies

64.99   Other financial service activities, except insurance and pension funding, n.e.c.

This class includes:

– other financial service activities primarily concerned with distributing funds other than by making loans:

64.99/1   Security dealing on own account

This subclass includes:

– dealing for own account by securities dealers
– activities of Stock Exchange money brokers
– activities of inter-dealer brokers
– dealing in financial futures, options and other derivatives for own account
– dealing in commodities for investment purposes

This subclass excludes:

– *security and other dealing on behalf of others, see 66.12*

64.99/2   Factoring

This subclass includes:

– activities of companies specialising in debt factoring and invoice discounting

This subclass excludes:

– *bill collection without debt buying up, see 82.91*

64.99/9   Other financial service activities, except insurance and pension funding, (not including security dealing on own account and factoring) n.e.c.

This subclass includes:

– activities of underwriters of stock and share issues
– securitisation activities
– writing of swaps, options and other hedging arrangements
– activities of viatical settlement companies

This subclass excludes:

– *activities of financial holding companies, see 64.20/5*
– *trade, leasing and renting of real estate property, see 68*
– *financial leasing, see 64.91*
– *management activities of non-financial holding companies, see 64.20*
– *grant-giving activities by membership organisations, see 94.99*

65   **Insurance, reinsurance and pension funding, except compulsory social security**

This division includes the underwriting of annuities and insurance policies and investing premiums to build up a portfolio of financial assets to be used against future claims. Provision of direct insurance and reinsurance are included.

65.1   **Insurance**

This group includes life insurance with or without a substantial savings element and non-life insurance.

65.11   **Life insurance**

This class includes:

– underwriting annuities and life insurance policies, disability income insurance policies, and accidental death and dismemberment insurance policies (with or without a substantial savings element)

65.12　　　　Non-life insurance

This class includes:

– provision of insurance services other than life insurance:
  ■ accident and fire insurance
  ■ health insurance
  ■ travel insurance
  ■ property insurance
  ■ motor, marine, aviation and transport insurance
  ■ pecuniary loss and liability insurance

65.2　　　　Reinsurance

65.20　　　　Reinsurance

65.20/1　　　Life reinsurance

This subclass includes the activities of assuming all or part of the risk associated with existing insurance policies originally underwritten by other insurance carriers.

This subclass includes only:

– those institutions authorised by the Financial Services Authority under the Insurance Companies Act 1982 and whose main activity is the carrying on of life re-insurance
– those institutions in the European Economic Area entitled to carry on insurance business in the UK on the basis of their home state authorisation and whose main activity is life re-insurance

65.20/2　　　Non-life reinsurance

This subclass includes the activities of assuming all or part of the risk associated with existing insurance policies originally underwritten by other insurance carriers.

This subclass includes:

– those institutions authorised by the Financial Services Authority under the Insurance Companies Act 1982 and whose main activity is the carrying on of non-life re-insurance business
– those institutions in the European Economic Area entitled to carry on insurance business in the UK on the basis of their home state authorisation and whose main activity is non-life re-insurance

65.3　　　　Pension funding

65.30　　　　Pension funding

This class includes legal entities (i.e. funds, plans and/or programmes) organised to provide retirement income benefits exclusively for the sponsor's employees or members. This includes pension plans with defined benefits, as well as individual plans where benefits are simply defined through the member's contribution.

This class includes:

– employee benefit plans
– pension funds and plans
– retirement plans

This class excludes:

– *management of pension funds, see 66.30*
– *compulsory social security schemes, see 84.30*

66　　　　　Activities auxiliary to financial services and insurance activities

This division includes the provision of services involved in or closely related to financial service activities, but not themselves providing financial services. The primary breakdown of this division is according to the type of financial transaction or funding served.

66.1　　　　Activities auxiliary to financial services, except insurance and pension funding

This group includes the furnishing of physical or electronic marketplaces for the purpose of facilitating the buying and selling of stocks, stock options, bonds or commodity contracts.

66.11　　　　Administration of financial markets

This class includes the operation and supervision of financial markets other than by public authorities, such as:

– commodity contracts exchanges
– futures commodity contracts exchanges

- securities exchanges
- stock exchanges
- stock or commodity options exchanges

66.12   **Security and commodity contracts brokerage**

This class includes:

- dealing in financial markets on behalf of others (e.g. stock broking) and related activities
- securities brokerage
- commodity contracts brokerage
- activities of bureaux de change etc.

This class excludes:

- *dealing in markets on own account, see 64.99*
- *portfolio management, on a fee or contract basis, see 66.30*

66.19   **Other activities auxiliary to financial services, except insurance and pension funding**

This class includes activities auxiliary to financial service activities n.e.c. such as:

- financial transaction processing and settlement activities, including for credit card transactions
- investment advisory services
- activities of mortgage advisers and brokers

This class also includes:

- trustee, fiduciary and custody services on a fee or contract basis

This class excludes:

- *activities of insurance agents and brokers, see 66.22*
- *management of funds, see 66.30*

66.2   **Activities auxiliary to insurance and pension funding**

This group includes acting as agents (i.e. brokers) in selling annuities and insurance policies or providing other employee benefits and insurance and pension related services such as claims adjustment and third party administration.

66.21   **Risk and damage evaluation**

This class includes the provision of administration services of insurance, such as:

- assessing insurance claims
  - claims adjusting
  - risk assessing
  - risk and damage evaluation
  - average and loss adjusting
- settling insurance claims

This class excludes:

- *appraisal of real estate, see 68.31*
- *appraisal for other purposes, see 74.90*
- *investigation activities, see 80.30*

66.22   **Activities of insurance agents and brokers**

This class includes:

- activities of insurance agents and brokers (insurance intermediaries) in selling, negotiating or soliciting of annuities and insurance and reinsurance policies

66.29   **Other activities auxiliary to insurance and pension funding**

This class includes:

- activities involved in or closely related to insurance and pension funding (except financial intermediation, claims adjusting and activities of insurance agents):
  - salvage administration
  - actuarial services

This class excludes:

- *marine salvage activities, see 52.22*

66.3　　　　　**Fund management activities**

66.30　　　　　Fund management activities

This class includes portfolio and fund management activities on a fee or contract basis, for individuals, businesses and others, such as:

- management of mutual funds
- management of other investment funds
- management of pension funds

K

Section L   **Real Estate Activities**

This section includes acting as lessors, agents and/or brokers in one or more of the following: selling or buying real estate, renting real estate, providing other real estate services such as appraising real estate or acting as real estate escrow agents. Activities in this section may be carried out on own or leased property and may be done on a fee or contract basis.

Also included is the building of structures, combined with maintaining ownership or leasing of such structures.

This section includes real estate property managers.

68   **Real estate activities**

68.1   **Buying and selling of own real estate**

68.10   **Buying and selling of own real estate**

This class includes:

– buying and selling of self-owned real estate:
  ■ apartment buildings and dwellings
  ■ non-residential buildings, including exhibition halls, self-storage facilities, malls and shopping centres
  ■ land

This class also includes:

– subdividing real estate into lots, without land improvement

This class excludes:

– *development of building projects for sale, see 41.10*
– *subdividing and improving of land, see 42.99*

68.2   **Renting and operating of own or leased real estate**

68.20   **Renting and operating of own or leased real estate**

68.20/1   **Renting and operating of Housing Association real estate**

This subclass includes:

– renting and operating of self-owned or leased real estate which is used for housing association activities only

This subclass also includes:

– renting and operating of housing association owned real estate:
  ■ apartment buildings and dwellings
  ■ land
– providing homes for rent for those who are vulnerable, on low income and/or key workers such as nurses
– sheltered or supported homes
– operation of services to help residents improve their communities
– offering homes through a shared ownership scheme
– development for building projects for own operation

68.20/2   **Letting and operating of conference and exhibition centres**

68.20/9   **Letting and operating of own or leased real estate (other than Housing Association real estate and conference and exhibition services) n.e.c.**

This subclass includes:

– renting and operating of self-owned or leased real estate:
  ■ apartment buildings and dwellings
  ■ non-residential buildings, excluding conference and exhibition centres, self-storage facilities
  ■ land
– providing of homes and furnished or unfurnished flats or apartments for more permanent use, typically on a monthly or annual basis

This subclass also includes:

– operation of residential mobile home sites
– development for building projects for own operation

This subclass excludes:

– *operation of hotels, suite hotels, holiday homes, rooming houses, campgrounds, trailer parks and other non-residential or short-stay accommodation places, see division 55*

68.3　　　　**Real estate activities on a fee or contract basis**

68.31　　　Real estate agencies

This class includes the provision of real estate activities by real estate agencies:

– intermediation in buying, selling and renting of real estate on a fee or contract basis
– advisory activities and appraisal services in connection with buying, selling and renting of real estate, on a fee or contract basis
– real estate escrow agents activities

This class excludes:

– *legal activities, see 69.10*

68.32　　　Management of real estate on a fee or contract basis

This class also includes:

– rent-collecting agencies

This class excludes:

– *legal activities, see 69.10*
– *facilities support services (combination of services such as general interior cleaning, maintenance and making minor repairs, trash disposal, guard and security), see 81.10*
– *management of facilities, such as military bases, prisons, and other facilities (except computer facilities management), see 81.10*

L

Section M   **Professional, Scientific and Technical Activities**

This section includes specialised professional, scientific and technical activities. These activities require a high degree of training, and make specialised knowledge and skills available to users.

69   **Legal and accounting activities**

This division includes legal representation of one party's interest against another party, whether or not before courts or other judicial bodies by, or under supervision of, persons who are members of the bar, such as advice and representation in civil cases, advice and representation in criminal actions, advice and representation in connection with labour disputes.

It also includes preparation of legal documents such as articles of incorporation, partnership agreements or similar documents in connection with company formation, patents and copyrights, preparation of deeds, wills, trusts, etc. as well as other activities of notaries public, civil law notaries, bailiffs, arbitrators, examiners and referees.

It also includes accounting and bookkeeping services such as auditing of accounting records, preparing financial statements and bookkeeping.

69.1   **Legal activities**

69.10   Legal activities

This class includes:

- legal representation of one party's interest against another party, whether or not before courts or other judicial bodies by, or under supervision of, persons who are members of the bar:
  - advice and representation in civil cases
  - advice and representation in criminal cases
  - advice and representation in connection with labour disputes
- general counselling and advising, preparation of legal documents:
  - articles of incorporation, partnership agreements or similar documents in connection with company formation
  - patents and copyrights
  - preparation of deeds, wills, trusts etc.
- other activities of notaries public, civil law notaries, bailiffs, arbitrators, examiners and referees

This class excludes:

- *law court activities, see 84.23*

69.10/1   Barristers at law

This subclass includes:

- members of the legal profession who have been called to the bar
- Advocates of the Scottish Bar

This subclass excludes:

- *law court activities, see 84.23*

69.10/2   Solicitors

This subclass includes:

- members of the legal profession qualified to deal with:
  - conveyancing
  - drawing up of wills
  - advising clients on legal matters
  - instructing barristers, etc.

69.10/9   Activities of patent and copyright agents; other legal activities (other than those of barristers and solicitors) n.e.c.

This subclass includes:

- preparation, drawing up and certification activities
- the provision of advice regarding patents and copyrights
- other legal activities not elsewhere classified
- activities of notaries, bailiffs, arbitrators, examiners and referees etc.

This subclass excludes:

- *law court activities, see 84.23*

69.2   **Accounting, bookkeeping and auditing activities; tax consultancy**

69.20   Accounting, bookkeeping and auditing activities; tax consultancy

69.20/1   Accounting, and auditing activities

This subclass includes:

– preparation or auditing of financial accounts
– examination of accounts and certification of their accuracy

This subclass excludes:

– *management consultancy on accounting systems, budgetary control procedures, see 70.22*
– *bill collection, see 82.91*

69.20/2   Bookkeeping activities

This subclass includes:

– recording of commercial transactions from businesses or others

This subclass excludes:

– *data-processing and tabulation activities, see 63.11*
– *management consultancy on accounting systems, budgetary control procedures, see 70.22*
– *bill collection, see 82.91*

69.20/3   Tax consultancy

This subclass includes:

– preparation of personal and business income tax returns
– advisory activities and representation on behalf of clients before tax authorities

This subclass excludes:

– *data-processing and tabulation activities, see 63.11*
– *management consultancy on accounting systems, budgetary control procedures, see 70.22*
– *bill collection, see 82.91*

70   **Activities of head offices; management consultancy activities**

This division includes the provision of advice and assistance to businesses and other organisations on management issues, such as strategic and organisational planning; financial planning and budgeting; marketing objectives and policies; human resource policies, practices, and planning; production scheduling; and control planning. It also includes the overseeing and managing of other units of the same company or enterprise, i.e. the activities of head offices.

70.1   **Activities of head offices**

70.10   Activities of head offices

This class includes the overseeing and managing of other units of the company or enterprise; undertaking the strategic or organisational planning and decision making role of the company or enterprise; exercising operational control and managing the day-to-day operations of their related units.

This class includes activities of:

– head offices
– centralised administrative offices
– corporate offices
– district and regional offices
– subsidiary management offices

This class excludes:

– *activities of holding companies, not engaged in managing, see 64.20*

70.2   **Management consultancy activities**

70.21   Public relations and communication activities

This class includes the provision of advice, guidance and operational assistance, including lobbying activities, to businesses and other organisations on public relations and communication.

This class excludes:

– *advertising agencies and media representation services, see 73.1*
– *market research and public opinion polling, see 73.20*

70.22   Business and other management consultancy activities

70.22/1   Financial management

This subclass includes the provision of advice, guidance and operational assistance to businesses and other organisations on issues such as cost reduction and other financial issues; compensation and retirement strategies.

This may include advice, guidance or operational assistance to businesses and the public service regarding:

- design of accounting methods or procedures, cost accounting programmes, budgetary control procedures

This subclass excludes:

- *design of computer software for accounting systems, see 62.01*
- *accounting, bookkeeping and auditing activities, tax consulting, see 69.20*

70.22/9   Management consultancy activities (other than financial management)

This subclass includes the provision of advice, guidance and operational assistance to businesses and other organisations on management issues, such as corporate strategic and organisational planning, business process re-engineering, change management, marketing objectives and policies; human resource policies, practices and planning; production scheduling and control planning.

This provision of business services may include advice, guidance or operational assistance to businesses and the public service regarding:

- advice and help to businesses and public services in planning, organisation, efficiency and control, management information etc.

This subclass excludes:

- *legal advice and representation, see 69.10*
- *architectural and engineering advisory activities, see 71.11, 71.12*
- *environmental, agronomy, security and similar consulting activities, see 74.90*
- *executive placement or search consulting activities, see 78.10*
- *educational consulting activities, see 85.60*

71   Architectural and engineering activities; technical testing and analysis

This division includes the provision of architectural services, engineering services, drafting services, building inspection services and surveying and mapping services. It also includes the performance of physical, chemical, and other analytical testing services.

71.1   Architectural and engineering activities and related technical consultancy

This group includes the provision of architectural services, engineering services, drafting services, building inspection services and surveying and mapping services and the like.

71.11   Architectural activities

71.11/1   Architectural activities

This subclass includes:

- architectural consulting activities:
  - building design and drafting
  - supervision of construction

This subclass excludes:

- *activities of computer consultants, see 62.02, 62.09*
- *interior decorating, see 74.10*

71.11/2   Urban planning and landscape architectural activities

This subclass includes:

- town and city planning and landscape architecture

71.12   Engineering activities and related technical consultancy

71.12/1   Engineering design activities for industrial process and production

This subclass comprises engineering design (i.e. applying physical laws and principles of engineering in the design of machines, materials, instruments, structures, processes and systems) for industrial process and production. It includes:

- drawing up of preliminary drafts, project development, specification of plans of execution or exact specifications on behalf of the contracting authority for the construction of industrial process and production

This subclass excludes:

- *industrial design, see 74.10*

71.12/2   Engineering related scientific and technical consulting activities

This subclass comprises the provision of surveying and mapping services and the like. It includes:

- geophysical, geologic and seismic surveying
- geological and prospecting activities:
  - surface measurements and observations designed to yield information on sub-surface structure and the location of petroleum, natural gas and mineral deposits and of ground water
- geodetic surveying activities:
  - land and boundary surveying activities
  - hydrologic surveying activities
  - subsurface surveying activities
  - cartographic and spatial information activities

This subclass excludes:

- *test drilling in connection with mining operations, see 09.10, 09.90*
- *development or publishing of associated software, see 58.29, 62.01*
- *activities of computer consultants, see 62.02, 62.09*
- *technical testing, see 71.20*
- *research and development activities related to engineering, see 72.19*
- *aerial photography, see 74.20*

71.12/9   Other engineering activities (not including engineering design for industrial process and production or engineering related scientific and technical consulting activities)

This subclass comprises the provision of architectural engineering services, drafting services, building inspection services and surveying and mapping services and the like. It includes:

- engineering design (i.e. applying physical laws and principles of engineering in the design of machines, materials, instruments, structures, processes and systems) and consulting activities for:
  - projects involving civil engineering, hydraulic engineering, traffic engineering
  - projects elaboration and realisation relative to electrical and electronic engineering, mining engineering, chemical engineering, mechanical, industrial and systems engineering, safety engineering
  - water management projects
- elaboration of projects using air conditioning, refrigeration, sanitary and pollution control engineering, acoustical engineering etc.
- machinery and industrial plan design, other than for industrial process and production

This subclass also includes:

- integrated engineering activities for turnkey projects

This subclass excludes:

- *engineering design activities for industrial process and production, see 71.12/1*

71.2   **Technical testing and analysis**

71.20   Technical testing and analysis

This class includes the performance of physical, chemical and other analytical testing of all types of materials and products, such as:

- acoustics and vibration testing
- testing of composition and purity of minerals etc.
- testing activities in the field of food hygiene, including veterinary testing and control in relation to food production
- testing of physical characteristics and performance of materials, such as strength, thickness, durability, radioactivity etc.
- qualification and reliability testing
- performance testing of complete machinery: motors, automobiles, electronic equipment etc.
- radiographic testing of welds and joints
- failure analysis
- testing and measuring of environmental indicators: air and water pollution etc.
- certification of products, including consumer goods, motor vehicles, aircraft, pressurised containers, nuclear plants etc.
- periodic road-safety testing of motor vehicles
- testing with use of models or mock-ups (e.g. of aircraft, ships, dams etc.)
- operation of police laboratories

This class excludes:

- *testing of animal specimens, see 75.00*
- *diagnostic imaging, testing and analysis of medical and dental specimens, see 86*

72   **Scientific research and development**

This division includes the activities of three types of research and development: 1) basic research: experimental or theoretical work undertaken primarily to acquire new knowledge of the underlying foundations of phenomena and observable facts, without particular application or use in view, 2) applied research: original investigation undertaken in order to acquire new knowledge, directed primarily towards a specific practical aim or objective and 3) experimental development: systematic work, drawing on existing knowledge gained from research and/or practical experience, directed to producing new materials, products and devices, to installing new processes, systems and services, and to improving substantially those already produced or installed.

Research and experimental development activities in this division are subdivided into two categories: natural sciences and engineering; social sciences and the humanities.

This division excludes:

– *market research, see 73.20*

72.1   **Research and experimental development on natural sciences and engineering**

This group comprises basic research, applied research, experimental development in natural sciences and engineering.

72.11   Research and experimental development on biotechnology

This class includes research and experimental development on biotechnology:

– DNA/RNA: genomics, pharmacogenomics, gene probes, genetic engineering, DNA/RNA sequencing/synthesis/amplification, gene expression profiling, and use of antisense technology
– proteins and other molecules: sequencing/synthesis/engineering of proteins and peptides (including large molecule hormones); improved delivery methods for large molecule drugs; proteomics, protein isolation and purification, signalling, identification of cell receptors
– cell and tissue culture and engineering: cell/tissue culture, tissue engineering (including tissue scaffolds and biomedical engineering), cellular fusion, vaccine/immune stimulants, embryo manipulation
– process biotechnology techniques: fermentation using bioreactors, bioprocessing, bioleaching,
– biopulping, biobleaching, biodesulphurisation, bioremediation, biofiltration and phytoremediation
– gene and RNA vectors: gene therapy, viral vectors
– bioinformatics: construction of databases on genomes, protein sequences; modelling complex biological processes, including systems biology
– nanobiotechnology: applies the tools and processes of nano/microfabrication to build devices for studying biosystems and applications in drug delivery, diagnostics etc.

72.19   Other research and experimental development on natural sciences and engineering

This class includes:

– research and experimental development on natural science and engineering other than biotechnological research and experimental development:
  ■ research and development on natural sciences
  ■ research and development on engineering and technology
  ■ research and development on medical sciences
  ■ research and development on agricultural sciences
  ■ interdisciplinary research and development, predominantly on natural sciences and engineering

72.2   **Research and experimental development on social sciences and humanities**

72.20   Research and experimental development on social sciences and humanities

This class includes:

– research and development on social sciences
– research and development on humanities
– interdisciplinary research and development, predominantly on social sciences and humanities

This class excludes:

– *market research, see 73.20*

73   **Advertising and market research**

This division includes the creation of advertising campaigns and placement of such advertising in periodicals, newspapers, radio and television, or other media as well as the design of display structures and sites.

73.1   **Advertising**

73.11   Advertising agencies

This class includes the provision of a full range of advertising services (i.e., through in-house capabilities or subcontracting), including advice, creative services, production of advertising material, and buying. It includes:

- creation and realisation of advertising campaigns:
  - creating and placing advertising in newspapers, periodicals, radio, television, the Internet and other media
  - creating and placing of outdoor advertising, e.g. billboards, panels, bulletins and frames, window dressing, showroom design, car and bus carding etc.
  - aerial advertising
  - distribution or delivery of advertising material or samples
  - creation of stands and other display structures and sites
- conducting marketing campaigns and other advertising services aimed at attracting and retaining customers
  - promotion of products
  - point-of-sale marketing
  - direct mail advertising
  - marketing consulting

This class excludes:

- *publishing of advertising material, see 58.19*
- *production of commercial messages for television and film, see 59.11*
- *production of commercial messages for radio, see 59.20*
- *market research, see 73.20*
- *advertising photography, see 74.20*
- *convention and trade show organisers, see 82.30*
- *mailing activities, see 82.19*

73.12　　Media representation

This class includes:

- media representation, i.e. sale or re-sale of time and space for various media soliciting advertising

This class excludes:

- *sale of advertising time or space directly by owners of the time or space (publishers etc.), see the corresponding activity class*
- *public-relations activities, see 70.21*

73.2　　**Market research and public opinion polling**

73.20　　Market research and public opinion polling

This class includes:

- investigation into market potential, awareness, acceptance and familiarity of goods and services and buying habits of consumers for the purpose of sales promotion and development of new goods and services, including statistical analyses of the results
- investigation into collective opinions of the public about political, economic and social issues and statistical analysis thereof

74　　**Other professional, scientific and technical activities**

This division includes the provision of professional scientific and technical services (except legal and accounting activities; architecture and engineering activities; technical testing and analysis; management and management consultancy activities; research and development and advertising activities).

74.1　　**Specialised design activities**

74.10　　Specialised design activities

This class includes:

- fashion design related to textiles, wearing apparel, shoes, jewellery, furniture and other interior decoration and other fashion goods as well as other personal or household goods
- industrial design, i.e. creating and developing designs and specifications that optimise the use, value and appearance of products, including the determination of the materials, mechanism, shape, colour and surface finishes of the product, taking into consideration human characteristics and needs, safety, market appeal in distribution, use and maintenance
- activities of graphic designers
- activities of interior decorators

This class excludes:

- *design and programming of web pages, see 62.01*
- *architectural design, see 71.11*
- *engineering design, i.e. applying physical laws and principles of engineering in the design of machines, materials, instruments, structures, processes and systems, see 71.12*

M

74.2   **Photographic activities**

74.20   Photographic activities

74.20/1   Portrait photographic activities

This subclass includes:

– portrait photography for passports, schools, weddings etc.

This subclass excludes:

– *operation of coin operated (self-service) photo machines, see 96.09*

74.20/2   Other specialist photography (not including portrait photography)

This subclass includes:

– aerial photography
– photographing of persons, objects or scenery using special apparatus and techniques e.g.
  ■ underwater photography
  ■ medical and biological photography
  ■ photomicrography
  ■ microfilming of documents

This subclass excludes:

– *cartographic and spatial information activities, see 71.12*

74.20/3   Film processing

This subclass includes:

– developing, printing and enlarging from client-taken negatives or cine-films
– film developing and photo printing laboratories
– one hour photo shops (not part of camera stores)
– mounting of slides
– copying and restoring or transparency retouching in connection with photographs

This subclass excludes:

– *processing motion picture film related to the motion picture and television industries, see 59.12*

74.20/9   Other photographic activities (not including portrait and other specialist photography and film processing) n.e.c.

This subclass includes:

– photography for commercials, publishers, fashion, real estate or tourism purposes
– videotaping of events: weddings, meetings etc.
– activities of photojournalists

74.3   **Translation and interpretation activities**

74.30   Translation and interpretation activities

74.9   **Other professional, scientific and technical activities n.e.c.**

74.90   Other professional, scientific and technical activities n.e.c.

This class includes a great variety of service activities generally delivered to commercial clients. It includes those activities for which more advanced professional, scientific and technical skill levels are required, but does not include ongoing, routine business functions that are generally of short duration.

74.90/1   Environmental consulting activities

This subclass includes:

– consulting activities for environmental projects

74.90/2   Quantity surveying activities

This subclass excludes:

– *research and development activities, see 72*

74.90/9   Other professional, scientific and technical activities (not including environmental consultancy or quantity surveying) n.e.c.

This subclass includes:

– business brokerage activities, i.e. arranging for the purchase and sale of small and medium-sized businesses, including professional practices, but not including real estate brokerage

- patent brokerage activities (arranging for the purchase and sale of patents)
- appraisal activities other than for real estate and insurance (for antiques, jewellery, etc.)
- bill auditing and freight rate information
- weather forecasting activities
- security consultants
- agronomy consulting
- other technical consulting
- activities of consultants other than architecture, engineering, environment and management consultants

This subclass also includes:

- activities carried out by agents and agencies on behalf of individuals usually involving the obtaining of engagements in motion picture, theatrical production or other entertainment or sports attractions and the placement of books, plays, artworks, photographs etc., with publishers, producers etc.

This subclass excludes:

- *wholesale of used motor vehicles by auctioning, see 45.1*
- *online auction activities (retail), see 47.91*
- *activities of auctioning houses (retail), see 47.79*
- *activities of real estate brokers, see 68.31*
- *activities of environmental consultants, see 74.90/1*
- *activities of quantity surveyors, see 74.90/2*
- *bookkeeping activities, see 69.20*
- *activities of management consultants, see 70.22*
- *activities of architecture and engineering consultants, see 71.1*
- *industrial and machinery design, see 71.12, 74.10*
- *veterinary testing and control in relation to food production, see 71.20*
- *display of advertisement and other advertising design, see 73.11*
- *creation of stands and other display structures and sites, see 73.11*
- *activities of convention and trade show organisers, see 82.30*
- *activities of independent auctioneers, see 82.99*
- *administration of loyalty programmes, see 82.99*
- *consumer credit and debt counselling, see 88.99*

M

75   Veterinary activities

This division includes the provision of animal health care and control activities for farm animals or pet animals. These activities are carried out by qualified veterinarians in veterinary hospitals as well as when visiting farms, kennels or homes, in own consulting and surgery rooms or elsewhere. It also includes animal ambulance activities.

75.0   Veterinary activities

75.00   Veterinary activities

This class includes:

- animal health care and control activities for farm animals
- animal health care and control activities for pet animals

These activities are carried out by qualified veterinarians when working in veterinary hospitals as well as when visiting farms, kennels or homes, in own consulting and surgery rooms or elsewhere.

This class also includes:

- activities of veterinary assistants or other auxiliary veterinary personnel
- clinico-pathological and other diagnostic activities pertaining to animals
- animal ambulance activities

This class excludes:

- *farm animal boarding activities without health care, see 01.62*
- *sheep shearing, see 01.62*
- *herd testing services, droving services, agistment activities, poultry caponising, see 01.62*
- *activities related to artificial insemination, see 01.62*
- *pet animal boarding activities without health care, see 96.09*

Section N   **Administrative and Support Service Activities**

This section includes a variety of activities that support general business operations. These activities differ from those in section M, in that their primary purpose is not the transfer of specialised knowledge.

77   **Rental and leasing activities**

This division includes the renting and leasing of tangible and non-financial intangible assets, including a wide array of tangible goods, such as automobiles, computers, consumer goods, and industrial machinery and equipment, to customers in return for a periodic rental or lease payment. It is subdivided into: (1) the renting of motor vehicles, (2) the renting of recreational and sports equipment and personal and household equipment, (3) the leasing of other machinery and equipment of the kind often used for business operations, including other transport equipment and (4) the leasing of intellectual property products and similar products.

Only the provision of operating leases is included in this division.

This division excludes:

– *financial leasing, see 64.91*
– *renting of real estate, see section L*
– *renting of equipment with operator, see corresponding classes according to activities carried out with this equipment, e.g. construction (section F), transportation (section H)*

77.1   **Renting and leasing of motor vehicles**

77.11   **Renting and leasing of cars and light motor vehicles**

This class includes:

– renting and operational leasing of the following types of vehicles:
   ■ passenger cars and other light motor vehicles (with a weight not exceeding 3.5 tonnes) without driver

This class excludes:

– *renting or leasing of cars or light motor vehicles with driver, see 49.32, 49.39*

N

77.12   **Renting and leasing of trucks**

This class includes:

– renting and operational leasing of the following types of vehicles:
   ■ trucks, utility trailers and heavy motor vehicles (with a weight exceeding 3.5 tonnes)
   ■ recreational vehicles

This class excludes:

– *renting or leasing of heavy goods vehicles or trucks with driver, see 49.41*

77.2   **Renting and leasing of personal and household goods**

This group includes the renting of personal and household goods as well as renting of recreational and sports equipment and video tapes. Activities generally include short-term renting of goods although in some instances, the goods may be leased for longer periods of time.

77.21   **Renting and leasing of recreational and sports goods**

This class includes renting of recreational and sports equipment:

– pleasure boats, canoes, sailboats
– bicycles
– beach chairs and umbrellas
– other sports equipment
– skis

This class excludes:

– *renting of pleasure boats and sailing boats with crew, see 50.10, 50.30*
– *renting of video tapes and disks, see 77.22*
– *renting of other personal and household goods n.e.c., see 77.29*
– *renting of leisure and pleasure equipment as an integral part of recreational facilities, see 93.29*

77.22   **Renting of video tapes and disks**

This class includes the renting of video tapes, records, CDs, DVDs etc.

77.29   **Renting and leasing of other personal and household goods**

77.29/1   Renting and leasing of media entertainment equipment

This subclass includes:

– renting of radios, televisions, video recorders, DVD players and the like

This subclass excludes:

– *renting of video tapes, records, CDs, DVDs etc., see 77.22*
– *renting of video tapes, records, CDs, DVDs etc. by libraries see 91.01*

77.29/9   Renting and leasing of other personal and household goods (other than media entertainment equipment)

This subclass includes:

– renting of all kinds of household or personal goods, to households or industries (except recreational and sports equipment):
  - textiles, wearing apparel and footwear
  - furniture, pottery and glass, kitchen and tableware, electrical appliances and house wares
  - jewellery, musical instruments, scenery and costumes
  - books, journals and magazines
  - machinery and equipment used by amateurs or as a hobby e.g. tools for home repairs
  - flowers and plants
  - electronic equipment, other than media entertainment equipment, for household use

This subclass class excludes:

– *renting of cars, trucks, trailers and recreational vehicles without driver, see 77.1*
– *renting of recreational and sports goods, see 77.21*
– *renting and leasing of media entertainment equipment, see 77.29/1*
– *renting of video tapes and disks, see 77.22*
– *renting of motorcycles and caravans without driver, see 77.39*
– *renting of office furniture, see 77.33*
– *provision of linen, work uniforms and related items by laundries, see 96.01*

77.3   **Renting and leasing of other machinery, equipment and tangible goods**

77.31   Renting and leasing of agricultural machinery and equipment

This class includes:

– renting and operational leasing of agricultural and forestry machinery and equipment without operator:
  - renting of products produced by class 28.30, such as agricultural tractors etc.

This class excludes:

– *renting of agricultural and forestry machinery or equipment with operator, see 01.61, 02.40*

77.32   Renting and leasing of construction and civil engineering machinery and equipment

This class includes:

– renting and operational leasing of construction and civil engineering machinery and equipment without operator:
  - crane lorries
  - scaffolds and work platforms, without erection and dismantling

This class excludes:

– *renting of construction and civil engineering machinery or equipment with operator, see division 43*

77.33   Renting and leasing of office machinery and equipment (including computers)

This class includes:

– renting and operational leasing of office machinery and equipment without operator:
  - computers and computer peripheral equipment
  - duplicating machines, typewriters and word-processing machines
  - accounting machinery and equipment: cash registers, electronic calculators etc.
  - office furniture

77.34   Renting and leasing of water transport equipment

77.34/1   Renting and leasing of passenger water transport equipment

This subclass includes:

– renting and operational leasing of passenger water-transport equipment, such as commercial boats and ships, without operator

This subclass excludes:

– *renting of water-transport equipment with operator, see division 50*
– *renting of pleasure boats, see 77.21*
– *financial leasing, see 64.91*

77.34/2   Renting and leasing of freight water transport equipment

This subclass includes:

– renting and operational leasing of freight water-transport equipment, such as commercial boats and ships, without operator

This subclass excludes:

– *renting of water-transport equipment with operator, see division 50*
– *financial leasing, see 64.91*

77.35   Renting and leasing of air transport equipment

77.35/1   Renting and leasing of passenger air transport equipment

This subclass includes:

– renting and operational leasing of passenger air transport equipment without operator

This subclass excludes:

– *renting of air-transport equipment with operator, see division 51*
– *financial leasing, see 64.91*

77.35/2   Renting and leasing of freight air transport equipment

This subclass includes:

– renting and operational leasing of freight air transport equipment without operator

This subclass excludes:

– *renting of air-transport equipment with operator, see division 51*
– *financial leasing, see 64.91*

77.39   Renting and leasing of other machinery, equipment and tangible goods n.e.c.

This class includes:

– renting and operational leasing, without operator, of other machinery and equipment that are generally used as capital goods by industries:
  ▪ engines and turbines
  ▪ machine tools
  ▪ mining and oilfield equipment
  ▪ professional radio, television and communication equipment
  ▪ motion picture production equipment
  ▪ measuring and controlling equipment
  ▪ other scientific, commercial and industrial machinery
– renting and operational leasing of land-transport equipment (other than motor vehicles) without drivers:
  ▪ motorcycles, caravans and campers etc.
  ▪ railway vehicles

This class also includes:

– renting of accommodation or office containers
– renting of animals (e.g. herds, race horses)
– renting of containers
– renting of pallets

This class excludes:

– *renting of bicycles, see 77.21*
– *renting of agricultural and forestry machinery and equipment, see 77.31*
– *renting of construction and civil engineering machinery and equipment, see 77.32*
– *renting of office machinery and equipment, including computers, see 77.33*

77.4　　　**Leasing of intellectual property and similar products, except copyrighted works**

77.40　　　Leasing of intellectual property and similar products, except copyrighted works

This class includes the activities of allowing others to use intellectual property products and similar products for which a royalty payment or licensing fee is paid to the owner of the product (i.e. the asset holder). The leasing of these products can take various forms, such as permission for reproduction, use in subsequent processes or products, operating businesses under a franchise etc. The current owners may or may not have created these products.

This class includes:

- leasing of intellectual property products (except copyrighted works, such as books or software)
- receiving royalties or licensing fees for the use of:
  - patented entities
  - trademarks or service marks
  - brand names
  - mineral exploration and evaluation
  - franchise agreements

This class excludes:

- *acquisition of rights and publishing, see divisions 58 and 59*
- *producing, reproducing and distributing copyrighted works (books, software, film), see divisions 58, 59*
- *leasing of real estate, see 68.20*
- *leasing of tangible products (assets), see groups 77.1, 77.2, 77.3*

78　　　**Employment activities**

This division includes activities of listing employment vacancies and referring or placing applicants for employment, where the individuals referred or placed are not employees of the employment agencies, supplying workers to clients' businesses for limited periods of time to supplement the workforce of the client, and the activities of providing other human resources.

This division includes:

- executive search and placement activities
- activities of theatrical casting agencies

This division excludes:

- *activities of agents for individual artists, see 74.90*

78.1　　　**Activities of employment placement agencies**

78.10　　　Activities of employment placement agencies

78.10/1　　　Motion picture, television and other theatrical casting

This subclass includes:

- activities of casting agencies and bureaus, such as theatrical casting agencies

This subclass excludes:

- *activities of personal theatrical or artistic agents or agencies, see 74.90*

78.10/9　　　Activities of employment placement agencies (other than motion picture, television and other theatrical casting) n.e.c.

This subclass includes listing employment vacancies and referring or placing applicants for employment, where the individuals referred or placed are not employees of the employment agencies.

This subclass includes:

- personnel search, selection referral and placement activities including executive placement and search activities
- activities of on-line employment placement agencies

This subclass excludes:

- *activities of personal theatrical or artistic agents or agencies, see 74.90*

78.2　　　**Temporary employment agency activities**

78.20　　　Temporary employment agency activities

This class includes the activities of supplying workers to clients' businesses for limited periods of time to temporarily replace or supplement the workforce of the client, where the individuals provided are employees of the temporary help service unit. However, units classified here do not provide direct supervision of their employees at the clients' work sites.

78.3   **Other human resources provision**

78.30   Human resources provision

This class includes the activities of providing human resources for client businesses. The units classified here represent the employer of record for the employees on matters relating to payroll, taxes, and other fiscal and human resource issues, but they are not responsible for direction and supervision of employees.

The provision of human resources is typically done on a long-term or permanent basis and the units classified here provide a wide range of human resource and personnel management duties associated with this provision.

This class excludes:

– *provision of human resources functions together with supervision or running of the business, see the class in the respective economic activity of that business*
– *provision of human resources to temporarily replace or supplement the workforce of a client, see 78.20*

79   **Travel agency, tour operator and other reservation service and related activities**

This division includes the activity of agencies, primarily engaged in selling travel, tour, transportation and accommodation services to the general public and commercial clients and the activity of arranging and assembling tours that are sold through travel agencies or directly by agents such as tour operators; and other travel-related services including reservation services. The activities of tourist guides and tourism promotion activities are also included.

79.1   **Travel agency and tour operator activities**

This group includes the activities of agencies, primarily engaged in selling travel, tour, transportation and accommodation services to the general public and commercial clients and the activity of arranging and assembling tours that are sold through travel agencies or directly by agents such as tour operators.

79.11   Travel agency activities

This class includes:

the activities of agencies, primarily engaged in selling travel, tour, transportation and accommodation services on a wholesale or retail basis to the general public and commercial clients.

79.12   Tour operator activities

This class includes:

– arranging and assembling tours that are sold through travel agencies or directly by tour operators. The tours may include any or all of the following:
  ■ transportation
  ■ accommodation
  ■ food
  ■ visits to museums, historical or cultural sites, theatrical, musical or sporting events

79.9   **Other reservation service and related activities**

79.90   Other reservation service and related activities

79.90/1   Activities of tourist guides

This subclass includes:

– tourist guide services

This subclass excludes:

– *the activities of travel agencies and tour operators, see classes 79.11 and 79.12*

79.90/9   Other reservation service and related activities (not including activities of tourist guides)

This subclass includes:

– other travel-related reservation services:

  ■ reservations for transportation, hotels, restaurants, car rentals, entertainment and sport etc.
– time-share exchange services
– ticket sales activities for theatrical, sports and other amusement and entertainment events
– visitor assistance services:
  ■ provision of travel information to visitors
– tourism promotion activities

This subclass excludes:

– *the activities of travel agencies and tour operators, see classes 79.11 and 79.12*
– *organisation and management of events such as meetings, conventions and conferences, see 82.30*

80 Security and investigation activities

This division includes security-related services such as: investigation and detective services; guard and patrol services; picking up and delivering money, receipts, or other valuable items with personnel and equipment to protect such properties while in transit; operation of electronic security alarm systems, such as burglar and fire alarms, where the activity focuses on remote monitoring these systems, but often involves also sale, installation and repair services. If the latter components are provided separately, they are excluded from this division and classified in retail sale, construction etc.

80.1 Private security activities

80.10 Private security activities

This class includes the provision of one or more of the following: guard and patrol services, picking up and delivering money, receipts, or other valuable items with personnel and equipment to protect such properties while in transit.

This class includes:

- armoured car services
- bodyguard services
- polygraph services
- fingerprinting services
- security guard services
- security shredding of information on any media

This class excludes:

- *public order and safety activities, see 84.24*

80.2 Security systems service activities

80.20 Security systems service activities

This class includes:

- monitoring or remote monitoring of electronic security alarm systems, such as burglar and fire alarms, including their installation and maintenance
- installing, repairing, rebuilding, and adjusting mechanical or electronic locking devices, safes and security vaults in connection with later monitoring and remote monitoring

The units carrying out these activities may also engage in selling such security systems, mechanical or electronic locking devices, safes and security vaults.

This class excludes:

- *installation of security systems, such as burglar and fire alarms, without later monitoring, see 43.21*
- *retail sale of electrical security alarm systems, mechanical or electronic locking devices, safes and security vaults in specialised stores, without monitoring, installation or maintenance services, see 47.59*
- *security consultants, see 74.90*
- *public order and safety activities, see 84.24*
- *providing key duplication services, see 95.29*

80.3 Investigation activities

80.30 Investigation activities

This class includes:

- investigation and detective service activities
- activities of all private investigators, independent of the type of client or purpose of investigation

81 Services to buildings and landscape activities

This division includes the provision of a number of general support services, such as the provision of a combination of support services within a client's facilities, the interior and exterior cleaning of buildings of all types, cleaning of industrial machinery, cleaning of trains, buses, planes, etc., cleaning of the inside of road and sea tankers, disinfecting and exterminating activities for buildings, ships, trains, etc., bottle cleaning, street sweeping, snow and ice removal, provision of landscape care and maintenance services and provision of these services along with the design of landscape plans and/or the construction (i.e. installation) of walkways, retaining walls, decks, fences, ponds, and similar structures.

81.1 Combined facilities support activities

81.10 Combined facilities support activities

This class includes the provision of a combination of support services within a client's facilities. These services include general interior cleaning, maintenance, trash disposal, guard and security, mail routing, reception, laundry and related services to support operations within facilities. These support activities are performed by operating staff who are not involved with or responsible for the core business or activities of the client.

This class excludes:

- *provision of only one of the support services (e.g. general interior cleaning services) or only a single function (e.g. heating), see the appropriate class according to the service provided*
- *provision of management and operating staff for the complete operation of a client's establishment, such as a hotel, restaurant, mine, or hospital, see the class of the unit operated*
- *provision of on site management and operation of a client's computer systems and/or data processing facilities, see 62.03*
- *operation of correctional facilities on a contract or fee basis, see 84.23*

81.2   **Cleaning activities**

This group includes the activities of general interior cleaning of all types of buildings, exterior cleaning of buildings, specialised cleaning activities for buildings or other specialised cleaning activities, cleaning of industrial machinery, cleaning of the inside of road and sea tankers, disinfecting and extermination activities for buildings and industrial machinery, bottle cleaning, street sweeping, snow and ice removal.

This group excludes:

- *agricultural pest control, see 01.61*
- *cleaning of new buildings immediately after construction, 43.39*
- *steam-cleaning, sand blasting and similar activities for building exteriors, see 43.99*
- *carpet and rug shampooing, drapery and curtain cleaning, see 96.01*

81.21   **General cleaning of buildings**

This class includes:

- general (non-specialised) cleaning activities of all types of buildings, such as:
  - offices
  - houses or apartments
  - factories
  - shops
  - institutions
- general (non-specialised) cleaning of other business and professional premises and multiunit residential buildings

These activities are mostly interior cleaning although they may include the cleaning of associated exterior areas such as windows or passageways.

This class excludes:

- *specialised cleaning activities, such as window cleaning, chimney cleaning, cleaning of fireplaces, stoves, furnaces, incinerators, boilers, ventilation ducts, exhaust units, see 81.22*

81.22   **Other building and industrial cleaning activities**

81.22/1   **Window cleaning services**

This subclass includes:

- cleaning of windows in dwellings and other buildings
- cleaning services for exterior windows using swing stages

81.22/2   **Specialised cleaning services**

This subclass includes:

- cleaning services for hospitals
- cleaning services for computer rooms
- specialised cleaning services of reservoirs and tanks, these being part of either industrial sites or transport equipment
- decontamination services
- cleaning services of heat and air-ducts
- sterilisation of objects or premises e.g. operating theatres

81.22/3   **Furnace and chimney cleaning services**

This subclass includes:

- furnace and chimney cleaning, cleaning of fireplaces, stoves, furnaces, incinerators and boilers

This subclass excludes:

- *maintenance services to central heating installations, see 33.11 and 43.22*

81.22/9   Building and industrial cleaning activities (other than window cleaning, specialised cleaning and furnace and chimney cleaning services) n.e.c.

This subclass includes the activities of:

- exterior cleaning of buildings of all types, including offices, factories, shops, institutions and other business and professional premises and multiunit residential buildings
- cleaning of industrial machinery
- other building and industrial cleaning activities, n.e.c.

This subclass excludes:

- *steam cleaning and blasting and similar activities for building exteriors, see 43.99*

81.29   Other cleaning services

81.29/1   Disinfecting and extermination services

This subclass includes:

- disinfecting of dwellings and other buildings
- exterminating insects, rodents and other pests
- fumigation services and pest control services

This subclass excludes:

- *agriculture pest control, see 01.61*
- *impregnation of timber see 16.10*

81.29/9   Cleaning services (other than disinfecting and extermination services) n.e.c.

This subclass includes the activities of:

- swimming pool cleaning and maintenance activities
- cleaning of trains, buses, planes, etc.
- cleaning of the inside of road and sea tankers
- bottle cleaning
- street sweeping and snow and ice removal
- rental of lavatory cubicles
- other cleaning activities, n.e.c.

This subclass excludes:

- *automobile cleaning, car wash, see 45.20*

81.3   **Landscape service activities**

81.30   Landscape service activities

This class includes planting, care and maintenance of:

- parks and gardens for:
  - private and public housing
  - public and semi-public buildings (schools, hospitals, administrative buildings, church buildings etc.)
  - municipal grounds (parks, green areas, cemeteries etc.)
  - highway greenery (roads, train lines and tramlines, waterways, ports)
  - industrial and commercial buildings
- greenery for:
  - buildings (roof gardens, façade greenery, indoor gardens etc.)
  - sports grounds (football fields, golf courses etc.), play grounds, lawns for sunbathing and other recreational parks
  - stationary and flowing water (basins, alternating wet areas, ponds, swimming pools, ditches, watercourses, plant sewage systems)
- plants for protection against noise, wind, erosion, visibility and dazzling

This class excludes:

- *commercial production and planting for commercial production of plants, trees, see divisions 01, 02*
- *tree nurseries and forest tree nurseries, see 01.30, 02.10*
- *keeping land in good environmental condition for agricultural use, see 01.61*
- *construction activities for landscaping purposes, see section F*
- *landscape design and architecture activities, see 71.11*

82 **Office administrative, office support and other business support activities**

This division includes the provision of a range of day to day office administrative services, as well as ongoing routine business support functions for others, on a contract or fee basis.

This division also includes all support service activities typically provided to businesses not elsewhere classified.

Units classified in this division do not provide operating staff to carry out the complete operations of a business.

82.1 **Office administrative and support activities**

This group includes the provision of a range of day to day office administrative services, such as financial planning, billing and record keeping, personnel and physical distribution and logistics for others on a contract or fee basis.

This group also includes support activities for others on a contract or fee basis, that are ongoing routine business support functions that businesses and organisations traditionally do for themselves.

Units classified in this group do not provide operating staff to carry out the complete operations of a business. Units engaged in one particular aspect of these activities are classified according to that particular activity.

82.11 Combined office administrative service activities

This class includes the provision of a combination of day to day office administrative services, such as reception, financial planning, billing and record keeping, personnel and mail services etc. for others on a contract or fee basis.

This class excludes:

- *provision of only one particular aspect of these activities, see class according to that particular activity*
- *provision of the operating staff without supervision, see 78*

82.19 Photocopying, document preparation and other specialised office support activities

This class includes a variety of copying, document preparation and specialised office support activities. The document copying/printing activities included here cover only short-run type printing activities.

This class includes:

- document preparation
- document editing or proofreading
- typing and word processing
- secretarial support services
- transcription of documents, and other secretarial services
- letter or resume writing
- provision of mailbox rental and other postal and mailing services, such as pre-sorting, addressing, etc.
- photocopying
- duplicating
- blueprinting
- other document copying services (without also providing printing services, such as offset printing, quick printing, digital printing, pre-press services)

This class excludes:

- *printing of documents (offset printing, quick printing etc.), see 18.12*
- *pre-press services, see 18.13*
- *developing and organising mail advertising campaigns, see 73.11*
- *specialised stenotype services such as court reporting, see 82.99*
- *public stenography services, see 82.99*

82.2 **Activities of call centres**

82.20 Activities of call centres

This class includes the activities of:

- inbound call centres, answering calls from clients by using human operators, automatic call distribution, computer telephone integration, interactive voice response systems or similar methods to receive orders, provide product information, deal with customer requests for assistance or address customer complaints
- outbound call centres using similar methods to sell or market goods or services to potential customers, undertake market research or public opinion polling and similar activities for clients

82.3 **Organisation of conventions and trade shows**

82.30 Organisation of conventions and trade shows

82.30/1   Activities of exhibition and fairorganisers

This subclass includes the organisation, promotion and/or management of events, such as business and trade shows, and conventions, whether or not including the management and provision of the staff to operate the facilities in which these events take place.

82.30/2   Activities of conference organisers

This subclass includes the organisation of conferences and meetings, whether or not including the management and provision of the staff to operate the facilities in which these events take place.

82.9   **Business support service activities n.e.c.**

This group includes the activities of collection agencies, credit bureaus and all support activities typically provided to businesses not elsewhere classified.

82.91   Activities of collection agencies and credit bureaus

82.91/1   Activities of collection agencies

This subclass includes:

– collection of payments for claims and remittance of payments collected to the clients, such as bill or debt collection services

82.91/2   Activities of credit bureaus

This subclass includes:

– compiling information, such as credit and employment histories on individuals and credit histories on businesses, and providing the information to financial institutions, retailers, and others who have a need to evaluate the creditworthiness of these persons and businesses

82.92   Packaging activities

This class includes:

– packaging activities on a fee or contract basis, whether or not these involve an automated process:
  ■ bottling of liquids, including beverages and food
  ■ packaging of solids (blister packaging, foil-covered etc.)
  ■ security packaging of pharmaceutical preparations
  ■ labelling, stamping and imprinting
  ■ parcel-packing and gift-wrapping

This class excludes:

– *manufacture of soft drinks and production of mineral water, see 11.07*
– *packaging activities incidental to transport, see 52.29*

82.99   Other business support service activities n.e.c.

This class includes:

– providing verbatim reporting and stenotype recording of live legal proceedings and transcribing subsequent recorded materials, such as:
  ■ court reporting or stenotype recording services
  ■ public stenography services
– real-time (i.e. simultaneous) closed captioning of live television performances of meetings, conferences
– address bar coding services
– bar code imprinting services
– fundraising organisation services on a contract or fee basis
– repossession services
– parking meter coin collection services
– activities of independent auctioneers
– administration of loyalty programmes
– other support activities typically provided to businesses not elsewhere classified

This class excludes:

– *provision of document transcription services, see 82.19*
– *providing film or tape captioning or subtitling services, see 59.12*

N

Section O   **Public Administration and Defence; Compulsory Social Security**

This section includes activities of a governmental nature, normally carried out by the public administration. This includes the enactment and judicial interpretation of laws and their pursuant regulation, as well as the administration of programmes based on them, legislative activities, taxation, national defence, public order and safety, immigration services, foreign affairs and the administration of government programmes.

This section also includes compulsory social security activities.

The legal or institutional status is not, in itself, the determining factor for an activity to belong in this section, rather than the activity being of a nature specified in the previous paragraph. This means that activities classified elsewhere in SIC do not fall under this section, even if carried out by public entities. For example, administration of the school system (i.e. regulations, checks, curricula) falls under this section, but teaching itself does not (see section P), and a prison or military hospital is classified to health (see section Q). Similarly, some activities described in this section may be carried out by non-government units.

84   **Public administration and defence; compulsory social security**

84.1   **Administration of the State and the economic and social policy of the community**

This group includes general administration (e.g. executive, legislative, financial administration etc. at all levels of government) and supervision in the field of social and economic life.

84.11   General public administration activities

This class includes:

– executive and legislative administration of central, regional and local bodies
– administration and supervision of fiscal affairs:
  - operation of taxation schemes
  - duty/tax collection on goods and tax violation investigation
  - customs administration
– budget implementation and management of public funds and public debt:
  - raising and receiving of money and control of their disbursement
– administration of overall (civil) research and development policy and associated funds
– administration and operation of overall economic and social planning and statistical services at the various levels of government

This class excludes:

– operation of government owned or occupied buildings, see 68.2, 68.3
– administration of research and development policies intended to increase personal well-being and of associated funds, see 84.12
– administration of research and development policies intended to improve economic performance and competitiveness, see 84.13
– administration of defence-related research and development policies and of associated funds, see 84.22
– operation of government archives, see 91.01

84.12   Regulation of the activities of providing health care, education, cultural services and other social services, excluding social security

This class includes:

– public administration of programmes aimed to increase personal well-being:
  - health
  - education
  - culture
  - sport
  - recreation
  - environment
  - housing
  - social services
– public administration of research and development policies and associated funds for these areas

This class also includes:

– sponsoring of recreational and cultural activities
– distribution of public grants to artists
– administration of potable water supply programmes
– administration of waste collection and disposal operations
– administration of environmental protection programmes
– administration of housing programmes

This class excludes:

- *sewage, refuse disposal and remediation activities, see divisions 37, 38, 39*
- *compulsory social security activities, see 84.30*
- *education activities, see section P*
- *human health-related activities, see division 86*
- *museums and other cultural institutions, see division 91*
- *activities of government operated libraries and archives, see 91.01*
- *sporting or other recreational activities, see division 93*

84.13　　Regulation of and contribution to more efficient operation of businesses

This class includes:

- public administration and regulation, including subsidy allocation, for different economic sectors:
  - agriculture
  - land use
  - energy and mining resources
  - infrastructure
  - transport
  - communication
  - hotels and tourism
  - wholesale and retail trade
- administration of research and development policies and associated funds to improve economic performance
- administration of general labour affairs
- implementation of regional development policy measures, e.g. to reduce unemployment

This class excludes:

- *research and experimental development activities, see division 72*

84.2　　Provision of services to the community as a whole

This group includes foreign affairs, defence and public order and safety activities.

84.21　　Foreign affairs

This class includes:

- administration and operation of the ministry of foreign affairs and diplomatic and consular missions stationed abroad or at offices of international organisations
- administration, operation and support for information and cultural services intended for distribution beyond national boundaries
- aid to foreign countries, whether or not routed through international organisations
- provision of military aid to foreign countries
- management of foreign trade, international financial and foreign technical affairs

This class excludes:

- *international disaster or conflict refugee services, see 88.99*

84.22　　Defence activities

This class includes:

- administration, supervision and operation of military defence affairs and land, sea, air and space defence forces such as:
  - combat forces of army, navy and air force
  - engineering, transport, communications, intelligence, material, personnel and other non-combat forces and commands
  - reserve and auxiliary forces of the defence establishment
  - military logistics (provision of equipment, structures, supplies etc.)
  - health activities for military personnel in the field
- administration, operation and support of civil defence forces
- support for the working out of contingency plans and the carrying out of exercises in which civilian institutions and populations are involved
- administration of defence-related research and development policies and related funds

This class excludes:

- *research and experimental development activities, see division 72*
- *provision of military aid to foreign countries, see 84.21*
- *activities of military tribunals, see 84.23*

– *provision of supplies for domestic emergency use in case of peacetime disasters, see 84.24*
– *educational activities of military schools, colleges and academies, see 85.4*
– *activities of military hospitals, see 86.10*

84.23   **Justice and judicial activities**

This class includes:

– administration and operation of administrative civil and criminal law courts, military tribunals and the judicial system, including legal representation and advice on behalf of the government or when provided by the government in cash or services
– rendering of judgements and interpretations of the law
– arbitration of civil actions
– prison administration and provision of correctional services, including rehabilitation services, regardless of whether their administration and operation is done by government units or by private units on a contract or fee basis

This class excludes:

– *advice and representation in civil, criminal and other cases, see 69.10*
– *activities of prison schools, see division 85*
– *activities of prison hospitals, see 86.10*

84.24   **Public order and safety activities**

This class includes:

– administration and operation of regular and auxiliary police forces supported by public authorities and of port, border, coastguards and other special police forces, including traffic regulation, alien registration, maintenance of arrest records
– provision of supplies for domestic emergency use in case of peacetime disasters

This class excludes:

– *operation of police laboratories, see 71.20*
– *administration and operation of military armed forces, see 84.22*

84.25   **Fire service activities**

This class includes:

– fire fighting and fire prevention:
    ■ administration and operation of regular and auxiliary fire brigades in fire prevention, fire fighting, rescue of persons and animals, assistance in civic disasters, floods, road accidents etc.

This class excludes:

– *forestry fire-protection and fire-fighting services, see 02.40*
– *oil and gas field fire fighting, see 09.10*
– *fire fighting and fire-prevention services at airports provided by non-specialised units, see 52.23*

84.3   **Compulsory social security activities**

84.30   Compulsory social security activities

This class includes:

– funding and administration of government-provided social security programmes:
    ■ sickness, work-accident and unemployment insurance
    ■ retirement pensions
    ■ programmes covering losses of income due to maternity, temporary disablement, widowhood etc.

This class excludes:

– *non-compulsory social security, see 65.30*
– *provision of welfare services and social work (without accommodation), see 88.10, 88.99*

Section P　　　**Education**

This section includes education at any level or for any profession. The instruction may be oral or written and may be provided by radio, television, Internet or via correspondence.

It includes education by the various institutions in the regular school system at its different levels as well as adult education, literacy programmes etc. Also included are military schools and academies, prison schools etc. at their respective levels. The section includes public as well as private education.

For each level of initial education, the classes include special education for physically or mentally disabled pupils.

The breakdown of the categories in this section is based on the level of education offered as defined by the levels of ISCED 1997. The activities of educational institutions providing courses on ISCED level 0 are classified in 85.10, on ISCED level 1 in 85.20, on ISCED levels 2-3 in group 85.3, on ISCED level 4 in 85.41 and on ISCED level 5-6 in 85.42.

This section also includes instruction primarily concerned with sport and recreational activities such as tennis or golf and education support activities.

85　　　**Education**

85.1　　　**Pre-primary education**

85.10　　　Pre-primary education

This class includes:

- pre-primary education (education preceding the first level). Pre-primary education is defined as the initial stage of organised instruction designed primarily to introduce very young children to a school-type environment, that is, to provide a bridge between the home and a school-based atmosphere.

This class excludes:

- *child day-care activities, see 88.91*

85.2　　　**Primary education**

85.20　　　Primary education

This class includes primary education: the furnishing of academic courses and associated course work that give students a sound basic education in reading, writing and mathematics and an elementary understanding of other subjects such as history, geography, natural science, social science, art and music.

Such education is generally provided for children, however the provision of literacy programmes within or outside the school system, which are similar in content to programmes in primary education but are intended for those considered too old to enter elementary schools, is also included (i.e. adult literacy programmes).

This class excludes:

- *adult education as defined in 85.5*
- *child day-care activities, including day nurseries for pupils, see 88.91*

85.3　　　**Secondary education**

This group includes the provision of general secondary and technical and vocational secondary education.

This group excludes:

- *adult education as defined in 85.5*

85.3　　　General secondary education

This class includes provision of the type of education that lays the foundation for lifelong learning and human development and is capable of furthering education opportunities. Such units provide programmes that are usually on a more subject-oriented pattern using more specialised teachers, and more often employ several teachers conducting classes in their field of specialisation.

Subject specialisation at this level often begins to have some influence even on the educational experience of those pursuing a general programme. Such programmes are designed to qualify students either for technical and vocational education or for entrance to higher education without any special subject prerequisite.

This class includes:

- lower general secondary education corresponding more or less to the period of compulsory school attendance
- upper general secondary education giving, in principle, access to higher education

85.32　　　Technical and vocational secondary education

This class includes provision of education typically emphasising subject matter specialisation and instruction in both theoretical background and practical skills generally associated with present or prospective employment. The aim of a programme can vary from preparation for a general field of employment to a very specific job.

221

This class includes:

– technical and vocational education below the level of higher education as defined in 85.4

This class also includes:

– tourist guide instruction
– instruction for chefs, hoteliers and restaurateurs
– cosmetology and barber schools
– computer repair training
– driving schools for occupational drivers e.g. of trucks, buses, coaches, schools for professional pilots

This class excludes:

– *technical and vocational higher education, see 85.4*
– *performing art instruction for recreation, hobby and self-development purposes, see 85.52*
– *automobile driving schools not intended for occupational drivers, see 85.53*
– *job training forming part of social work activities without accommodation, see 88.10, 88.99*

85.4   Higher education

This group includes the furnishing of post-secondary non-tertiary and academic courses and granting of degrees at baccalaureate, graduate or post-graduate level. The requirement for admission is a diploma at least at upper secondary education level.

This group excludes:

– *adult education as defined in 85.5*

85.41   Post-secondary non-tertiary education

This class includes provision of post-secondary education, which cannot be considered tertiary education. For example provision of supplementary post-secondary education to prepare for tertiary education or post-secondary non-tertiary vocational.

85.42   Tertiary education

This class includes:

– first, second and third stages of tertiary education

This class also includes:

– performing arts schools providing tertiary education

85.42/1   First-degree level higher education

This subclass includes:

– education of the type that leads to an award of first degree or equivalent either at university or other institute which provides such study

This subclass also includes:

– education leading to some professional qualifications
– study leading to a one year Post Graduate Certificate of Education (PGCE)

85.42/2   Post-graduate level higher education

This subclass includes:

– education that leads to a post-graduate degree or equivalent (e.g. Ph.D., MA, M.Sc.) either at university or other institute which provides such study

85.5   Other education

This group includes general continuing education and continuing vocational education and training for any profession, hobby or self-development purposes.

It includes camps and schools offering instruction in athletic activities to groups or individuals, foreign language instruction, instruction in the arts, drama or music or other instruction or specialised training, not comparable to the education in groups 85.1–85.4.

It excludes educational activities as outlined in groups 85.1-85.4 – i.e. pre-primary education, primary education, secondary education or higher education.

85.51   Sports and recreation education

This class includes the provision of instruction in athletic activities to groups of individuals, such as by camps and schools. Overnight and day sports instruction camps are also included. It does not include academic schools, colleges and universities.

Instruction may be provided in diverse settings, such as the unit's or client's training facilities, educational institutions or by other means. Instruction provided in this class is formally organised.

This class includes:

- sports instruction (baseball, basketball, cricket, football, etc.)
- camps, sports instruction
- gymnastics instruction
- riding instruction, academies or schools
- swimming instruction
- professional sports instructors, teachers, coaches
- martial arts instruction
- card game instruction (such as bridge)
- yoga instruction

This class excludes:

- *cultural education, see 85.52*

85.52　Cultural education

This class includes provision of instruction in the arts, drama and music. Units giving this type of instructions might be named "schools", "studios", "classes" etc. They provide formally organised instruction, mainly for hobby, recreational or self-development purposes, but such instruction does not lead to a professional diploma, baccalaureate or graduate degree.

This class includes:

- piano teachers and other music instruction
- art instruction
- dance instruction and dance studios
- drama schools (except academic)
- fine arts schools (except academic)
- performing arts schools (except academic)
- photography schools (except commercial)

This class excludes:

- *foreign language instruction, see 85.59*

85.53　Driving school activities

This class also includes:

- flying, sailing, shipping schools not issuing commercial certificates and permits

This class excludes:

- *driving schools for occupational drivers, see 85.32*

85.59　Other education n.e.c.

This class includes:

- education that is not definable by level
- academic tutoring
- learning centres offering remedial courses
- professional examination review courses
- language instruction and conversational skills instruction
- computer training
- religious instruction

This class also includes:

- lifeguard training
- survival training
- public speaking training
- speed reading instruction

This class excludes:

- *adult literacy programmes see 85.20*
- *general secondary education, see 85.31*

    – *technical and vocational secondary education, see 85.32*

    – *higher education, see 85.4*

85.6   **Educational support activities**

85.60   Educational support activities

This class includes:

– provision of non-instructional activities that support educational processes or systems:
  - educational consulting
  - educational guidance counselling activities
  - educational testing evaluation activities
  - educational testing activities
  - organisation of student exchange programmes

This class excludes:

– *research and experimental development on social sciences and humanities, see 72.20*

P

Section Q　　　Human Health and Social Work Activities

This section includes the provision of health and social work activities. It covers a wide range of activities, from health care provided by trained medical professionals in hospitals and other facilities, to residential care activities that still involve a degree of health care activities and to social work activities not involving the services of health care professionals.

86　　　　　　　Human health activities

This division includes activities of short or long-term hospitals, general or specialist medical, surgical, psychiatric and substance abuse hospitals, sanatoria, preventoria, medical nursing homes, asylums, mental hospitals, rehabilitation centres, leprosaria and other human health institutions which have accommodation facilities and which engage in providing diagnostic and medical treatment to inpatients with any of a wide variety of medical conditions.

It also includes medical consultation and treatment in the field of general and specialised medicine by general practitioners and medical specialists and surgeons. It includes dental practice activities of a general or specialised nature and orthodontic activities. Additionally, this division includes activities for human health not performed by hospitals or by practicing medical doctors but by paramedical practitioners legally recognised to treat patients.

86.1　　　　　Hospital activities

86.10　　　　　Hospital activities

86.10/1　　　　Hospital activities

This subclass includes:

- short- or long-term hospital activities, i.e. medical, diagnostic and treatment activities, of general hospitals (e.g. community and regional hospitals, hospitals of non-profit organisations, university hospitals, military-base and prison hospitals) and specialised hospitals (e.g. mental health and substance abuse hospitals, hospitals for infectious diseases, maternity hospitals, specialised sanatoriums)

The activities are chiefly directed to inpatients, are carried out under the direct supervision of medical doctors and include:

- services of medical and paramedical staff
- services of laboratory and technical facilities, including radiologic and anaesthesiologic services
- emergency room services
- provision of operating room services, pharmacy services, food and other hospital services
- services of family planning centres providing medical treatment such as sterilisation and termination of pregnancy, with accommodation

This subclass excludes:

- *laboratory testing and inspection of all types of materials and products, except medical, see 71.20*
- *veterinary activities, see 75.00*
- *health activities for military personnel in the field, see 84.22*
- *dental practice activities of a general or specialised nature, e.g. dentistry, endodontic and pediatric dentistry; oral pathology, orthodontic activities, see 86.23*
- *private consultants' services to inpatients, see 86.2*
- *medical laboratory testing, see 86.90*
- *ambulance transport activities, see 86.90*
- *medical nursing homes, see 86.10/2*

86.10/2　　　　Medical nursing home activities

The activities carried out here are chiefly directed to inpatients and carried out under the direct supervision of medical doctors.

This subclass excludes:

- *homes for the elderly, see 87.30*

86.2　　　　　Medical and dental practice activities

This group includes medical consultation and treatment provided by general medical practitioners and medical specialists, including surgeons, dentists etc.

These activities can be carried out in private practice, group practices and in hospital outpatient clinics, and in clinics such as those attached to firms, schools, homes for the aged, labour organisations and fraternal organisations, as well as in patients' homes.

This group also includes:

- private consultants' services to inpatients

86.21　　　　　General medical practice activities

This class includes:

- medical consultation and treatment in the field of general medicine carried out by general practitioners.

This class excludes:

- *inpatient hospital activities, see 86.10*
- *paramedical activities such as those of midwives, nurses and physiotherapists, see 86.90*

86.22　　Specialist medical practice activities

This class includes:

- medical consultation and treatment in the field of specialised medicine by medical specialists and surgeons

This class also includes:

- family planning centres providing medical treatment such as sterilisation and termination of pregnancy, without accommodation

This class excludes:

- *inpatient hospital activities, see 86.10*
- *activities of midwives, physiotherapists and other paramedical practitioners, see 86.90*

86.23　　Dental practice activities

This class includes:

- dental practice activities of a general or specialised nature, e.g. dentistry, endodontic and pediatric dentistry; oral pathology
- orthodontic activities

This class also includes:

- dental activities in operating rooms

This class excludes:

- *production of artificial teeth, denture and prosthetic appliances by dental laboratories, see 32.50*
- *inpatient hospital activities, see 86.10*
- *activities of dental paramedical personnel such as dental hygienists, see 86.90*

86.9　　Other human health activities

86.90　　Other human health activities

This class includes:

- activities for human health not performed by hospitals or by medical doctors or dentists:
    - activities of nurses, midwives, physiotherapists or other paramedical practitioners in the field of optometry, hydrotherapy, medical massage, occupational therapy, speech therapy, chiropody, homeopathy, chiropractic, acupuncture etc.

These activities may be carried out in health clinics such as those attached to firms, schools, homes for the elderly, labour organisations and fraternal organisations and in residential health facilities other than hospitals, as well as in own consulting rooms, patients' homes or elsewhere.

This class also includes:

- activities of dental paramedical personnel such as dental therapists, school dental nurses and dental hygienists, who may work remote from, but are periodically supervised by, the dentist
- activities of medical laboratories such as:
    - X-ray laboratories and other diagnostic imaging centres
    - blood analysis laboratories
- activities of blood banks, sperm banks, transplant organ banks etc.
- ambulance transport of patients by any mode of transport including aeroplanes.

These services are often provided during a medical emergency.

This class excludes:

- *production of artificial teeth, denture and prosthetic appliances by dental laboratories, see 32.50*
- *transfer of patients, with neither equipment for lifesaving nor medical personnel, see divisions 49, 50, 51*
- *non-medical laboratory testing, see 71.20*
- *testing activities in the field of food hygiene, see 71.20*
- *hospital activities, see 86.10*
- *medical and dental practice activities, see 86.2*
- *residential nursing care facilities, see 87.10*

87   **Residential care activities**

This division includes the provision of residential care combined with either nursing, supervisory or other types of care as required by the residents. Facilities are a significant part of the production process and the care provided is a mix of health and social services with the health services being largely some level of nursing services.

87.1   **Residential nursing care activities**

87.10   Residential nursing care activities

This class includes:

– activities of:
  ■ homes for the elderly with nursing care
  ■ convalescent homes
  ■ rest homes with nursing care
  ■ nursing care facilities
  ■ nursing homes

This class excludes:

– *in-home services provided by health care professionals, see division 86*

– *activities of homes for the elderly without or with minimal nursing care, see 87.30*

– *social work activities with accommodation, such as children's boarding homes and hostels, temporary homeless shelters, see 87.90*

87.2   **Residential care activities for learning disabilities, mental health and substance abuse**

87.20   Residential care activities for learning disabilities, mental health and substance abuse

This class includes the provision of residential care (but not licensed hospital care) to people with learning disabilities, mental illness, or substance abuse problems. Facilities provide room, board, protective supervision and counselling and some health care.

This class includes:

– activities of:
  ■ facilities for alcoholism or drug addiction treatment
  ■ psychiatric convalescent homes
  ■ residential group homes for the emotionally disturbed
  ■ learning disabilities facilities
  ■ mental health halfway houses

It also includes provision of residential care and treatment for patients with mental health and substance abuse illnesses.

This class excludes:

– *mental hospitals, see 86.10*

– *social work activities with accommodation, such as temporary homeless shelters, see 87.90*

87.3   **Residential care activities for the elderly and disabled**

87.30   Residential care activities for the elderly and disabled

This class includes the provision of residential and personal care services for the elderly and disabled who are unable to fully care for themselves and/or who do not desire to live independently. The care typically includes room, board, supervision, and assistance in daily living, such as housekeeping services. In some instances these units provide skilled nursing care for residents in separate on-site facilities.

This class includes:

– activities of:
  ■ assisted-living facilities
  ■ continuing care retirement communities
  ■ homes for the elderly with minimal nursing care
  ■ rest homes without nursing care

This class excludes:

– *activities of homes for the elderly with nursing care, see 87.10*

– *social work activities with accommodation where medical treatment or education are not important elements, see 87.90*

87.9   **Other residential care activities**

87.90   Other residential care activities

This class includes the provision of residential and personal care services for persons, except the elderly and disabled, who are unable to fully care for themselves or who do not desire to live independently.

This class includes:

– activities provided on a round-the-clock basis directed to provide social assistance to children and special categories of persons with some limits on ability for self-care, but where medical treatment or education are not important elements:
  ■ orphanages
  ■ children's boarding homes and hostels
  ■ temporary homeless shelters
  ■ institutions that take care of unmarried mothers and their children

The activities may be carried out by government offices or private organisations.

This class also includes:

– activities of:
  ■ halfway group homes for persons with social or personal problems
  ■ halfway homes for delinquents and offenders
  ■ juvenile correction homes

This class excludes:

– *funding and administration of compulsory social security programmes, see 84.30*
– *activities of nursing care facilities, see 87.10*
– *residential care activities for the elderly or disabled, see 87.30*
– *adoption activities, see 88.99*
– *short-term shelter activities for disaster victims, see 88.99*

88   **Social work activities without accommodation**

This division includes the provision of a variety of social assistance services directly to clients. The activities in this division do not include accommodation services, except on a temporary basis.

88.1   **Social work activities without accommodation for the elderly and disabled**

88.10   Social work activities without accommodation for the elderly and disabled

This class includes:

– social, counselling, welfare, referral and similar services which are aimed at the elderly and disabled in their homes or elsewhere and carried out by government offices or by private organisations, national or local self-help organisations and by specialists providing counselling services:
  ■ visiting of the elderly and disabled
  ■ day-care activities for the elderly or for disabled adults
  ■ vocational rehabilitation and habilitation activities for disabled persons provided that the education component is limited

This class excludes:

– *funding and administration of compulsory social security programmes, see 84.30*
– *activities similar to those described in this class, but including accommodation, see 87.30*
– *day-care activities for disabled children, see 88.91*

88.9   **Other social work activities without accommodation**

88.91   Child day-care activities

This class includes also:

– activities of day nurseries for pupils, including day-care activities for disabled children

88.99   Other social work activities without accommodation n.e.c.

This class includes:

– social, counselling, welfare, refugee, referral and similar services which are delivered to individuals and families in their homes or elsewhere and carried out by government offices or by private organisations, disaster relief organisations and national or local self-help organisations and by specialists providing counselling services:
  ■ welfare and guidance activities for children and adolescents
  ■ adoption activities, activities for the prevention of cruelty to children and others
  ■ household budget counselling, marriage and family guidance, credit and debt counselling services

- community and neighbourhood activities
- activities for disaster victims, refugees, immigrants etc., including temporary or extended shelter for them
- vocational rehabilitation and habilitation activities for unemployed persons provided that the education component is limited
- eligibility determination in connection with welfare aid, rent supplements or food stamps
- day facilities for the homeless and other socially weak groups
- charitable activities like fund-raising or other supporting activities aimed at social work

This class excludes:

- *funding and administration of compulsory social security programmes, see 84.30*
- *activities similar to those described in this class, but including accommodation, see 87.90*

Q

Section R   **Arts, Entertainment and Recreation**

This section includes a wide range of activities catering for various cultural, entertainment and recreational interests of the general public, including live performances, operation of museum sites, gambling, sports and recreation activities.

90   **Creative, arts and entertainment activities**

This division includes the operation of facilities and provision of services to meet the cultural and entertainment interests of the general public. This includes the production and promotion of, and participation in, live performances, events or exhibits intended for public viewing; the provision of artistic, creative or technical skills for the production of artistic products and live performances.

This division excludes:

- *the operation of museums of all kinds, botanical and zoological gardens; the preservation of historical sites; and nature reserves activities, see division 91*
- *gambling and betting activities, see division 92*
- *sports and amusement and recreation activities, see division 93*

Some units that provide cultural, entertainment or recreational facilities and services are classified in other divisions, such as:

- motion picture and video production and distribution, see 59.11, 59.12, 59.13
- motion picture projection, see 59.14
- radio and television broadcasting, see 60.1, 60.2

90.0   **Creative, arts and entertainment activities**

This group includes activities in the creative and performing arts and related activities.

90.01   Performing arts

This class includes:

- live theatrical presentations, concerts and opera or dance productions and other stage productions:
  - activities of groups, circuses or companies, orchestras or bands
  - activities of individual artists such as actors, dancers, musicians, lecturers or speakers

This class excludes:

- *activities of personal theatrical or artistic agents or agencies, see 74.90*
- *casting activities, see 78.10*

90.02   Support activities to performing arts

This class includes:

- support activities to performing arts for production of live theatrical presentations, concerts and opera or dance productions and other stage productions:
  - activities of directors, producers, stage-set designers and builders, scene shifters, lighting engineers etc.

This class also includes:

- activities of producers or entrepreneurs of arts live events, with or without facilities

This class excludes:

- *activities of personal theatrical or artistic agents or agencies, see 74.90*
- *casting activities, see 78.10*

90.03   Artistic creation

This class includes:

- activities of individual artists such as sculptors, painters, cartoonists, engravers, etchers etc.
- activities of individual writers, for all subjects including fictional writing, technical writing etc.
- activities of independent journalists
- restoring of works of art such as paintings etc.

This class excludes:

- *manufacture of statues, other than artistic originals, see 23.70*
- *restoring of organs and other historical musical instruments, see 33.19*
- *motion picture and video production, see 59.11, 59.12*
- *restoring of furniture (except museum type restoration), see 95.24*

R

90.04   Operation of arts facilities

This class includes:

– operation of concert and theatre halls and other arts facilities

This class excludes:

– *operation of cinemas, see 59.14*
– *activities of ticket agencies, see 79.90*
– *operation of museums of all kinds, see 91.02*

91   **Libraries, archives, museums and other cultural activities**

This division includes the activities of libraries and archives; the operation of museums of all kinds, botanical and zoological gardens; the operation of historical sites; nature reserves activities. It also includes the preservation and exhibition of objects, sites and natural wonders of historical, cultural or educational interest (e.g. world heritage sites, etc.).

This division excludes:

– *sports and amusement and recreation activities such as the operation of bathing beaches and recreation parks, see division 93*

91.0   **Libraries, archives, museums and other cultural activities**

91.01   Library and archive activities

This class includes:

– documentation and information activities of libraries of all kinds, reading, listening and viewing rooms, public archives providing service to the general public or to a special clientele, such as students, scientists, staff, members as well as operation of government archives:
   ■ organisation of a collection, whether specialised or not
   ■ cataloguing collections
   ■ lending and storage of books, maps, periodicals, films, records, tapes, works of art etc.
   ■ retrieval activities in order to comply with information requests etc.
– stock photo and movie libraries and services

91.01/1   Library activities

This subclass includes:

– selection, acquisition and organisation of a collection which may be specialist or general, for lending or reference
– cataloguing and preservation of collections
– lending and storage of books, periodicals, CDs, DVDs, maps, music etc including materials in alternative formats
– offering access to IT facilities including Internet
– retrieval activities in order to comply with information requests, reference enquiries etc, in person, via telephone, letter or Internet
– making services available to general public or particular groups of users of the organisation e.g., students, members, researchers, Members of Parliament
– additional services such as training courses (IT, information literacy, basic skills), user groups (reading, family history etc)
– organisation and running activities outside the library building for groups of users e.g. schools, conferences

91.01/2   Archive activities

This subclass includes:

– archive activities are about managing a unique accumulation of records (which can be paper, parchment, audio visual, digital) representing the transactions of an individual, family, organisation (including records of government, courts, businesses, charities) or society
– appraising and selecting archives to be kept for posterity on the basis of their continuing value as primary source material documenting the culture and history of an individual, organisation and society
– preserving/conserving archive records (ranging in age from 1000 years old to 21st century emails) protecting them from the main threats (poor environmental conditions, theft)
– making available archives of all kinds to general public or to a special clientele such as students, scientists, through providing physical access in search rooms or remote access via the Internet
– documenting archives according to the way they were organically created maintaining the original order, links between records and evidence of provenance
– managing and implementing information policy (such as Freedom of Information, Data Protection, and enabling citizens to exercise their rights to access information), and managing current records

91.02   Museum activities

This class includes:

– operation of museums of all kinds:
  - art museums, museums of jewellery, furniture, costumes, ceramics, silverware
  - natural history, science and technological museums, historical museums, including military museums
  - other specialised museums
  - open-air museums

This class excludes:

– *activities of commercial art galleries, see 47.78*
– *restoration of works of art and museum collection objects, see 90.03*
– *activities of libraries and archives, see 91.01*

91.03   Operation of historical sites and buildings and similar visitor attractions

This class includes:

– operation and preservation of historical sites and buildings

This class excludes:

– *renovation and restoration of historical sites and buildings, see section F*

91.04   Botanical and zoological gardens and nature reserve activities

This class includes:

– operation of botanical and zoological gardens, including children's zoos
– operation of nature reserves, including wildlife preservation, etc.

This class excludes:

– *landscape and gardening activities, see 81.30*
– *operation of sport fishing and hunting preserves, see 93.19*

92   **Gambling and betting activities**

This division includes the operation of gambling facilities such as casinos, bingo halls and video gaming terminals and the provision of gambling services, such as lotteries and off-track betting.

92.0   **Gambling and betting activities**

92.00   Gambling and betting activities

This class includes gambling and betting activities such as:

– sale of lottery tickets
– operation (exploitation) of coin-operated gambling machines
– operation of virtual gambling web sites
– bookmaking and other betting operations
– off-track betting
– operation of casinos, including "floating casinos"

93   **Sports activities and amusement and recreation activities**

This division includes the provision of recreational, amusement and sports activities (except museum activities, preservation of historical sites, botanical and zoological gardens and nature reserve activities; and gambling and betting activities).

Excluded from this division are dramatic arts, music and other arts and entertainment such as the production of live theatrical presentations, concerts and opera or dance productions and other stage productions, see division 90.

93.1   **Sports activities**

This group includes the operation of sports facilities; activities of sports teams or clubs primarily participating in live sports events before a paying audience; independent athletes engaged in participating in live sporting or racing events before a paying audience; owners of racing participants such as cars, dogs, horses, etc. primarily engaged in entering them in racing events or other spectator sports events; sports trainers providing specialised services to support participants in sports events or competitions; operators of arenas and stadiums; other activities of organising, promoting or managing sports events, n.e.c.

93.11   Operation of sports facilities

This class includes:

– the operation of facilities for indoor or outdoor sports events (open, closed or covered, with or without spectator seating):
  - football, hockey, cricket, rugby stadiums

- racetracks for car, dog, horse races
- swimming pools and stadiums
- track and field stadiums
- winter sports arenas and stadiums
- ice-hockey arenas
- boxing arenas
- golf courses
- bowling lanes

- organisation and operation of outdoor or indoor sports events for professionals or amateurs by organisations with own facilities

This class includes managing and providing the staff to operate these facilities.

This class excludes:

- *operation of ski lifts, see 49.39*
- *renting of recreation and sports equipment, see 77.21*
- *activities of fitness facilities, see 93.13*
- *park and beach activities, see 93.29*

93.12   **Activities of sport clubs**

This class includes the activities of sports clubs, which, whether professional, semi-professional or amateur clubs, give their members the opportunity to engage in sporting activities.

This class includes:

- the operation of sports clubs:
  - football clubs
  - bowling clubs
  - swimming clubs
  - golf clubs
  - boxing clubs
  - winter sports clubs
  - chess clubs
  - track and field clubs
  - shooting clubs, etc.

This class excludes:

- *sports instruction by individual teachers, trainers, see 85.51*
- *operation of sports facilities, see 93.11*
- *organisation and operation of outdoor or indoor sports events for professionals or amateurs by sports clubs with their own facilities, see 93.11*

93.13   **Fitness facilities**

This class includes:

- fitness and body-building clubs and facilities

This class excludes:

- *sports instruction by individual teachers, trainers, see 85.51*

93.19   **Other sports activities**

93.19/1   **Activities of racehorse owners**

This subclass includes:

- the seeking of sponsorship, appearance money and prize money

This subclass excludes:

- *activities of racing stables, see 93.19/9*
- *activities of riding academies, see 85.51*

93.19/9   **Other sports activities (not including activities of racehorse owners) n.e.c.**

This subclass includes:

- activities of producers or promoters of sports events, with or without facilities
- activities of individual own-account sportsmen and athletes, referees, judges, timekeepers etc.

- activities of sports leagues and regulating bodies
- activities related to promotion of sporting events
- activities of racing stables, kennels and garages
- operation of sport fishing and hunting preserves
- support activities for sport or recreational hunting and fishing
- activities of mountain guides

This subclass excludes:

- *activities of racehorse owners, see 93.19/1*
- *renting of sports equipment, see 77.21*
- *activities of sport and game schools, see 85.51*
- *activities of sports instructors, teachers, coaches, see 85.51*
- *activities of riding academies, see 85.51*
- *organisation and operation of outdoor or indoor sports events for professionals or amateurs by sports clubs with/without own facilities, see 93.11, 93.12*
- *park and beach activities, see 93.29*

93.2   **Amusement and recreation activities**

This group includes a wide range of units that operate facilities or provide services to meet the varied recreational interests of their patrons. It includes the operation of a variety of attractions, such as mechanical rides, water rides, games, shows, theme exhibits and picnic grounds.

This group excludes sports activities and dramatic arts, music and other arts and entertainment.

93.21   **Activities of amusement parks and theme parks**

This class includes activities of amusement parks or theme parks. It includes the operation of a variety of attractions, such as mechanical rides, water rides, games, shows, theme exhibits and picnic grounds.

93.29   **Other amusement and recreation activities**

This class includes activities related to entertainment and recreation (except amusement parks and theme parks) not elsewhere classified:

- operation (exploitation) of coin-operated games
- activities of recreation parks (without accommodation)
- operation of recreational transport facilities, e.g. marinas
- operation of ski hills
- renting of leisure and pleasure equipment as an integral part of recreational facilities
- fairs and shows of a recreational nature
- activities of beaches, including renting of facilities such as bathhouses, lockers, chairs etc.
- operation of dance floors

This class also includes activities of producers or entrepreneurs of live events other than arts or sports events, with or without facilities.

This class excludes:

- *operation of teleferics, funiculars, ski and cable lifts, see 49.39*
- *fishing cruises, see 50.10, 50.30*
- *provision of space and facilities for short stay by visitors in recreational parks and forests and campgrounds, see 55.30*
- *trailer parks, recreational camps, hunting and fishing camps, campsites and campgrounds, see 55.30*
- *beverage serving activities of discotheques, see 56.30*
- *theatrical and circus groups, see 90.01*

Section S   **Other Service Activities**

This section (as a residual category) includes the activities of membership organisations, the repair of computers and personal and household goods and a variety of personal service activities not covered elsewhere in the classification.

94   **Activities of membership organisations**

This division includes activities of organisations representing interests of special groups or promoting ideas to the general public. These organisations usually have a constituency of members, but their activities may involve and benefit non-members as well. The primary breakdown of this division is determined by the purpose that these organisations serve, namely interests of employers, self-employed individuals and the scientific community (group 94.1), interests of employees (group 94.2) or promotion of religious, political, cultural, educational or recreational ideas and activities (group 94.9).

94.1   **Activities of business, employers and professional membership organisations**

This group includes the activities of units that promote the interests of the members of business and employers organisations. In the case of professional membership organisations, it also includes the activities of promoting the professional interests of members of the profession.

94.11   Activities of business and employers membership organisations

This class includes:

- activities of organisations whose members' interests centre on the development and prosperity of enterprises in a particular line of business or trade, including farming, or on the economic growth and climate of a particular geographical area or political subdivision without regard for the line of business
- activities of federations of such associations
- activities of chambers of commerce, guilds and similar organisations
- dissemination of information, representation before government agencies, public relations and labour
- negotiations of business and employer organisations

This class excludes:

- *activities of trade unions, see 94.20*

94.12   Activities of professional membership organisations

This class includes:

- activities of organisations whose members' interests centre chiefly on a particular scholarly discipline or professional practice or technical field, such as medical associations, legal associations, accounting associations, engineering associations, architects associations etc.
- activities of associations of specialists engaged in scientific, academic or cultural activities, such as associations of writers, painters, performers of various kinds, journalists etc.
- dissemination of information, the establishment and supervision of standards of practice, representation
- before government agencies and public relations of professional organisations

This class also includes:

- activities of learned societies

This class excludes:

- *education provided by these organisations, see division 85*

94.2   **Activities of trade unions**

94.20   Activities of trade unions

This class includes:

- promoting of the interests of organised labour and union employees

This class also includes:

- activities of associations whose members are employees interested chiefly in the representation of their views concerning the salary and work situation, and in concerted action through organisation
- activities of single plant unions, of unions composed of affiliated branches and of labour organisations composed of affiliated unions on the basis of trade, region, organisational structure or other criteria

This class excludes:

- *education provided by such organisations, see division 85*

94.9   **Activities of other membership organisations**

This group includes the activities of units (except business and employers organisations, professional organisations, trade unions) that promote the interests of their members.

94.91   Activities of religious organisations

This class includes:

– activities of religious organisations or individuals providing services directly to worshippers in churches, mosques, temples, synagogues or other places
– activities of organisations furnishing monastery and convent services
– religious retreat activities

This class also includes:

– religious funeral service activities

This class excludes:

– *education provided by such organisations, see division 85*
– *health activities by such organisations, see division 86*
– *social work activities by such organisations, see divisions 87, 88*

94.92   Activities of political organisations

This class includes:

– activities of political organisations and auxiliary organisations such as young people's auxiliaries associated with a political party. These organisations chiefly engage in influencing decision-taking in public governing bodies by placing members of the party or those sympathetic to the party in political office and involve the dissemination of information, public relations, fund-raising etc.

94.99   Activities of other membership organisations n.e.c.

This class includes:

– activities of organisations not directly affiliated to a political party furthering a public cause or issue by means of public education, political influence, fund-raising etc.:
  ■ citizens initiative or protest movements
  ■ environmental and ecological movements
  ■ organisations supporting community and educational facilities n.e.c.
  ■ organisations for the protection and betterment of special groups, e.g. ethnic and minority groups
  ■ associations for patriotic purposes, including war veterans' associations
– consumer associations
– automobile associations
– associations for the purpose of social acquaintanceship such as rotary clubs, lodges etc.
– associations of youth, young persons' associations, student associations, clubs and fraternities etc.
– associations for the pursuit of a cultural or recreational activity or hobby (other than sports or games), e.g. poetry, literature and book clubs, historical clubs, gardening clubs, film and photo clubs, music and art clubs, craft and collectors' clubs, social clubs, carnival clubs etc.
– associations for the protection of animals

This class also includes:

– grant giving activities by membership organisations or others

This class excludes:

– *charitable activities like fund-raising aimed at social work, see 88.99*
– *activities of professional artistic groups or organisations, see 90.0*
– *activities of sports clubs, see 93.12*
– *activities of professional associations, see 94.12*

95   **Repair of computers and personal and household goods**

This division includes the repair and maintenance of computers peripheral equipment such as desktops, laptops, computer terminals, storage devices and printers.

It also includes the repair of communications equipment such as fax machines, two-way radios and consumer electronics such as radios and TVs, home and garden equipment such as lawn-mowers and blowers, footwear and leather goods, furniture and home furnishings, clothing and clothing accessories, sporting goods, musical instruments, hobby articles and other personal and household goods.

Excluded from this division is the repair of medical and diagnostic imaging equipment, measuring and surveying instruments, laboratory instruments, radar and sonar equipment, see 33.13.

95.1   **Repair of computers and communication equipment**

This group includes the repair and maintenance of computers and peripheral equipment and communications equipment.

95.11　　　Repair of computers and peripheral equipment

This class includes the repair of electronic equipment, such as computers and computing machinery and peripheral equipment.

This class includes the repair and maintenance of:

- desktop computers
- laptop computers
- magnetic disk drives, flash drives and other storage devices
- optical disk drives (CD-RW, CD-ROM, DVD-ROM, DVD-RW)
- printers
- monitors
- keyboards
- mice, joysticks, and trackball accessories
- internal and external computer modems
- dedicated computer terminals
- computer servers
- scanners, including bar code scanners
- smart card readers
- virtual reality helmets
- computer projectors

This class also includes the repair and maintenance of:

- computer terminals like automatic teller machines (ATM's); point-of-sale (POS) terminals, not mechanically operated
- hand-held computers (PDA's)

This class excludes:

- *the repair and maintenance of carrier equipment modems, see 95.12*

95.12　　　Repair of communication equipment

This class includes repair and maintenance of communications equipment such as:

- cordless telephones
- cellular phones
- carrier equipment modems
- fax machines
- communications transmission equipment (e.g. routers, bridges, modems)
- two-way radios
- commercial TV and video cameras

95.2　　　**Repair of personal and household goods**

This group includes the repair and servicing of personal and household goods.

95.21　　　Repair of consumer electronics

This class includes repair and maintenance of consumer electronics:

- repair of consumer electronics:
  - television, radio receivers
  - video cassette recorders (VCR)
  - CD players
  - household-type video cameras

95.22　　　Repair of household appliances and home and garden equipment

This class includes the repair and servicing household appliances and home and garden equipment:

- repair and servicing of household appliances:
  - refrigerators, stoves, washing machines, clothes dryers, room air conditioners, etc.
- repair and servicing of home and garden equipment:
  - lawnmowers, edgers, snow- and leaf-blowers, trimmers, etc.

This class excludes:

- *repair of hand held power tools, see 33.12*
- *repair of central air conditioning systems, see 43.22*

95.23   **Repair of footwear and leather goods**

This class includes repair and maintenance of footwear and leather goods:

- repair of boots, shoes, luggage and the like
- fitting of heels

95.24   **Repair of furniture and home furnishings**

This class includes:

- reupholstering, refinishing, repairing and restoring of furniture and home furnishings including office furniture

95.25   **Repair of watches, clocks and jewellery**

This class includes:

- repair of watches, clocks and their parts such as watch cases and housings of all materials; movements, chronometers, etc.
- repair of jewellery

This class excludes:

- *industrial engraving of metals, see 25.61*
- *repair of time clocks, time/date stamps, time locks and similar time recording devices, see 33.13*

95.29   **Repair of other personal and household goods**

This class includes repair of personal and household goods:

- repair of bicycles
- repair and alteration of clothing
- repair of sporting goods (except sporting guns) and camping equipment
- repair of books
- repair of musical instruments (except organs and historical musical instruments)
- repair of toys and similar articles
- repair of other personal and household goods
- piano-tuning

This class excludes:

- *industrial engraving of metals, see 25.61*
- *repair of hand held power tools, see 33.12*
- *repair of sporting and recreational guns, 33.11*

96   **Other personal service activities**

This division includes all service activities not mentioned elsewhere in the classification. Notably it includes types of services such as washing and (dry-)cleaning of textiles and fur products, hairdressing and other beauty treatment, funeral and related activities.

96.0   **Other personal service activities**

96.01   **Washing and (dry-)cleaning of textile and fur products**

This class includes:

- laundering and dry cleaning, pressing etc., of all kinds of clothing (including fur) and textiles, provided by mechanical equipment, by hand or by self-service coin-operated machines, whether for the general public or for industrial or commercial clients
- laundry collection and delivery
- carpet and rug shampooing and drapery and curtain cleaning, whether on clients' premises or not
- provision of linens, work uniforms and related items by laundries
- diaper supply services

This class excludes:

- *renting of clothing other than work uniforms, even if cleaning of these goods is an integral part of the activity, see 77.29*
- *repair and alteration of clothing etc., see 95.29*

96.02   **Hairdressing and other beauty treatment**

This class includes:

- hair washing, trimming and cutting, setting, dyeing, tinting, waving, straightening and similar activities for men and women
- shaving and beard trimming
- facial massage, manicure and pedicure, make-up etc.

This class excludes:

– *manufacture of wigs, see 32.99*

96.03　　**Funeral and related activities**

This class includes:

– burial and incineration of human or animal corpses and related activities:
  - preparing the dead for burial or cremation and embalming and morticians' services
  - providing burial or cremation services
  - renting of equipped space in funeral parlours
– renting or sale of graves
– maintenance of graves and mausoleums

This class excludes:

– *religious funeral service activities, see 94.91*
– *cemetery gardening, see 81.30*

96.04　　**Physical well-being activities**

This class includes:

– activities of Turkish baths, sauna and steam baths, solariums, reducing and slimming salons, massage salons etc.

This class excludes:

– *medical massage and therapy, see 86.90*
– *activities of health, fitness and body-building clubs and facilities, see 93.13*

96.09　　**Other personal service activities n.e.c.**

This class includes:

– astrological and spiritualists' activities
– social activities such as escort services, dating services, services of marriage bureaux
– pet care services such as boarding, grooming, sitting and training pets
– genealogical organisations
– activities of tattooing and piercing studios
– shoe shiners, porters, valet car parkers etc.
– concession operation of coin-operated personal service machines (photo booths, weighing machines, machines for checking blood pressure, coin-operated lockers etc.)

This class excludes:

– *veterinary activities, see 75.00*
– *coin-operated washing machines, see 96.01*
– *coin-operated gambling machines, see 92.00*

S

Section T	**Activities of Households as Employers; Undifferentiated Goods- and Services-Producing Activities of Households for Own Use**
97	**Activities of households as employers of domestic personnel**
97.0	**Activities of households as employers of domestic personnel**
97.00	Activities of households as employers of domestic personnel

This class includes the activities of households as employers of domestic personnel such as maids, cooks, waiters, valets, butlers, laundresses, gardeners, gatekeepers, stable-lads, chauffeurs, caretakers, governesses, babysitters, tutors, secretaries etc.

It allows the domestic personnel employed to state the activity of their employer in censuses or studies, even though the employer is an individual. The product produced by this activity is consumed by the employing household.

This class excludes:

– *provision of services such as cooking, gardening etc. by independent service providers (companies or individuals), see according to type of service*

98	**Undifferentiated goods- and services-producing activities of private households for own use**

This division includes the undifferentiated subsistence goods-producing and services-producing activities of households.

Households should be classified here only if it is impossible to identify a primary activity for the subsistence activities of the household. If the household engages in market activities, it should be classified according to the primary market activity carried out.

98.1	**Undifferentiated goods-producing activities of private households for own use**
98.10	Undifferentiated goods-producing activities of private households for own use

This class includes the undifferentiated subsistence goods-producing activities of households, that is to say, the activities of households that are engaged in a variety of activities that produce goods for their own subsistence. These activities include hunting and gathering, farming, the production of shelter and clothing and other goods produced by the household for its own subsistence.

If households are also engaged in the production of marketed goods, they are classified to the appropriate goods-producing industry of SIC. If they are principally engaged in a specific goods-producing subsistence activity, they are classified to the appropriate goods-producing industry of SIC.

98.2	**Undifferentiated service-producing activities of private households for own use**
98.20	Undifferentiated service-producing activities of private households for own use

This class includes the undifferentiated subsistence services-producing activities of households. These activities include cooking, teaching, caring for household members and other services produced by the household for its own subsistence.

If households are also engaged in the production of multiple goods for subsistence purposes, they are classified to the undifferentiated goods-producing subsistence activities of households.

T

Section U　　　**Activities of Extraterritorial Organisations and Bodies**

See class 9900

99　　　**Activities of extraterritorial organisations and bodies**

99.0　　　**Activities of extraterritorial organisations and bodies**

99.00　　　Activities of extraterritorial organisations and bodies

This class includes:

– activities of international organisations such as the United Nations and the specialised agencies of the United Nations system, regional bodies etc., the International monetary Fund, the World Bank, the World Customs Organisation, the Organisation for Economic Co-operation and Development, the Organisation of Petroleum Exporting Countries, the European Communities, the European Free Trade Association etc.

This class also includes:

– activities of diplomatic and consular missions when being determined by the country of their location rather than by the country they represent

U